Building Smart Homes with Raspberry Pi Zero

Build revolutionary and incredibly useful home automation projects with the all-new Pi Zero

Marco Schwartz

BIRMINGHAM - MUMBAI

Building Smart Homes with Raspberry Pi Zero

First published: October 2016

Production reference: 1241016

Published by Packt Publishing Ltd.
Livery Place
35 Livery Street
Birmingham B3 2PB, UK.

ISBN 978-1-78646-695-2

www.packtpub.com

Credits

Author
Marco Schwartz

Reviewer
Vasilis Tzivaras

Commissioning Editor
Kartikey Pandey

Acquisition Editor
Prachi Bisht

Content Development Editor
Trusha Shriyan

Technical Editors
Nirant Carvalho

Naveenkumar Jain

Copy Editors
Safis Editing

Sneha Singh

Project Coordinator
Kinjal Bari

Proofreader
Safis Editing

Indexer
Pratik Shirodkar

Graphics
Kirk D'Penha

Production Coordinator
Shantanu N Zagade

Cover Work
Shantanu N. Zagade

About the Author

Marco Schwartz is an electrical engineer, entrepreneur, and blogger. He has a master's degree in electrical engineering and computer science from Supélec, France, and a master's degree in micro engineering from the Ecole Polytechnique Fédérale de Lausanne (EPFL), Switzerland.

He has more than five years of experience working in the domain of electrical engineering. Marco's interests gravitate around electronics, home automation, the Arduino and Raspberry Pi platforms, open source hardware projects, and 3D printing.

He has several websites about Arduino, including the Open Home Automation website, which is dedicated to building home automation systems using open source hardware.

Marco has written another book on home automation and Arduino, called *Home Automation With Arduino: Automate Your Home Using Open-source Hardware*. He has also written a book on how to build Internet of Things projects with Arduino, called *Internet of Things with the Arduino Yun*, by Packt Publishing.

About the Reviewer

Vasilis Tzivaras is a software developer and hardware engineer who lives in Ioannina, Greece. He is currently an undergraduate student of the department of computer science and engineering at Ioannina. Along with his studies, he is working on many projects relevant to robotics, such as drones, home automation, and smart home systems using Arduino and the Raspberry Pi. He is also enthusiastic about clean energy solutions and cultural innovation ideas.

He has worked for the University Hospital of Ioannina as an assistant for various computer issues and has been a part of the support team of his CSE department for over a year. He has participated in IEEE UOI Student Branch and other big organizations, such as FOSSCOMM, with personal presentations for website designing, programming, Linux systems, and drones.

He is the chair of IEEE University of Ioannina Student Branch and has proposed many projects and solutions to automate homes and many other life problems by reducing the time of everyday routines. In addition to this, he has come up with ideas to entertain kids with funny and magical projects using Arduino-like hardware and open source software. Many of the projects can be found on his GitHub account under the name of BillyTziv.

Apart from *Building Smart Homes with Raspberry Pi Zero* and *Internet of Things with Arduino Cookbook*, he has also published a book named *Building a Quadcopter with Arduino*. He has also worked on another book *Programming in C*, which is not yet published. In addition to this, he has written for blogs, forums, guides, and small chapters, explaining and sharing his knowledge of computers, networks, and programming.

www.PacktPub.com

eBooks, discount offers, and more

Did you know that Packt offers eBook versions of every book published, with PDF and ePub files available? You can upgrade to the eBook version at `www.PacktPub.com` and as a print book customer, you are entitled to a discount on the eBook copy. Get in touch with us at `customercare@packtpub.com` for more details.

At `www.PacktPub.com`, you can also read a collection of free technical articles, sign up for a range of free newsletters and receive exclusive discounts and offers on Packt books and eBooks.

`https://www.packtpub.com/mapt`

Get the most in-demand software skills with Mapt. Mapt gives you full access to all Packt books and video courses, as well as industry-leading tools to help you plan your personal development and advance your career.

Why Subscribe?

- Fully searchable across every book published by Packt
- Copy and paste, print, and bookmark content
- On demand and accessible via a web browser

Table of Contents

Preface

The Raspberry Pi is an amazing development platform that was introduced back in 2012, along with the release of the first board. However, due to its price, it was not convenient for people to use it for smart home applications, where you need to deploy several modules at different places of your home. Usually, people building smart homes with this board used it as a central hub.

However, in 2016 the Raspberry Pi foundation released the Zero board, which is a smaller and much cheaper version of the Raspberry Pi board, and this changed everything for the home automation field. Now, you can actually use several of these boards in a home automation system and enjoy all the power and flexibility of the Raspberry Pi for all your projects.

This is exactly what I will teach you to do in this book. You will learn how to use the Raspberry Pi Zero board in several home automation projects, in order for you to build a smart home that is really tailored to your needs.

What this book covers

Chapter 1, *Configuring Your Raspberry Pi Zero Board*, teaches you how to get started with your Raspberry Pi Zero board and also install everything that you need to carry out all the projects that you will find in this book.

Chapter 2, *Measure Data Using Your Raspberry Pi Zero Board*, teaches you how to measure data from a sensor using the Raspberry Pi Zero board. You will also learn how to measure data from the sensor, store this data on the Pi, and plot it graphically.

Chapter 3, *Building a Smart Home Thermostat*, gets you right into the core topic of this book, that is, building your first home automation system. In this chapter, we will build a simple thermostat that will allow you to regulate the temperature in your home.

Chapter 4, Control Appliances from the Raspberry Pi Zero, shows you how to use the Raspberry Pi Zero board to easily control home appliances. As an example, we'll see how to control a DC motor and switch on/off appliances, such as lamps.

Chapter 5, Making a Smart Plug with the Raspberry Pi Zero, teaches you how to build your own version of a smart wireless plug that you can buy in a shop. We'll see how to build the same using the Raspberry Pi Zero board and how to customize it for your own needs.

Chapter 6, Sending Notifications using Raspberry Pi Zero, shows you how to send automated notifications from your Pi, for example to indicate that the temperature in your home is getting low. As examples, we'll see how to send text, email, and push notifications.

Chapter 7, Use the Raspberry Pi Zero to Build a Security System, shows you how to start integrating everything we saw so far in the book and build a security system using what we have learned so far. You'll, for example, learn how to transform your Raspberry Pi Zero board into a wireless security camera.

Chapter 8, Monitor Your Home from the Cloud, guides you through an amazing field: the Internet of Things. You will learn how to use the Internet of Things for your smart home and monitor it from anywhere in the world.

Chapter 9, Control Appliances from Anywhere, guides you into the field of the Internet of Things, this time by teaching you how to control home appliances from outside of your Wi-Fi network.

Chapter 10, Building a Home Automation System with Raspberry Pi Zero Boards, uses everything you learned in the book to build a complete home automation system for your entire home.

What you need for this book

For this book, the main component you will need is, of course, a Raspberry Pi Zero board. In the first chapter of the book, I will show you how to completely configure the board, so you can use it for the projects of this book. We will use some basic components at the start, such as sensors, and then move on to using more complex components. For every project, I will of course guide you step-by-step into building the hardware part so that you are not left behind.

On the software side, it is good if you actually have some existing programming skills, especially in JavaScript and in the Node.js framework. However, I will explain all the parts of each software piece of this book; so even if you don't have good programming skills in JavaScript you will be able to follow along.

Who this book is for

This book is for all the people who want to automate their homes and make it smarter, while at the same time having complete control on what they are doing. If that's your case, you will learn everything there is to learn in this book, on how to use the amazing Raspberry Pi Zero board to automate your home.

This book is also for makers who have played in the past with other development boards, such as Arduino. If that's your case, you will learn how to use the power of the Raspberry Pi platform to build smart homes. You will also learn to create projects that can't easily be done with other platforms, such as creating a wireless security camera with the Pi Zero.

Conventions

In this book, you will find a number of text styles that distinguish between different kinds of information. Here are some examples of these styles and an explanation of their meaning.

Code words in text, database table names, folder names, filenames, file extensions, pathnames, dummy URLs, user input, and Twitter handles are shown as follows: "You can now simply navigate to the IP address of the computer or Pi on which the application is running, followed by port `3000`."

A block of code is set as follows:

```
var request = require('request');
var sensorLib = require('node-dht-sensor');
```

Any command-line input or output is written as follows:

```
sudo npm install express request
```

New terms and **important words** are shown in bold. Words that you see on the screen, for example, in menus or dialog boxes, appear in the text like this: "You can now just click on **Stream** to access the live stream from the camera."

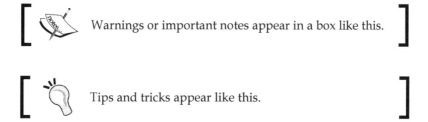

> Warnings or important notes appear in a box like this.

> Tips and tricks appear like this.

Reader feedback

Feedback from our readers is always welcome. Let us know what you think about this book—what you liked or disliked. Reader feedback is important for us as it helps us develop titles that you will really get the most out of.

To send us general feedback, simply e-mail feedback@packtpub.com, and mention the book's title in the subject of your message.

If there is a topic that you have expertise in and you are interested in either writing or contributing to a book, see our author guide at www.packtpub.com/authors.

Customer support

Now that you are the proud owner of a Packt book, we have a number of things to help you to get the most from your purchase.

Downloading the example code

You can download the example code files for this book from your account at http://www.packtpub.com. If you purchased this book elsewhere, you can visit http://www.packtpub.com/support and register to have the files e-mailed directly to you.

You can download the code files by following these steps:

1. Log in or register to our website using your e-mail address and password.
2. Hover the mouse pointer on the **SUPPORT** tab at the top.
3. Click on **Code Downloads & Errata**.
4. Enter the name of the book in the **Search** box.
5. Select the book for which you're looking to download the code files.
6. Choose from the drop-down menu where you purchased this book from.
7. Click on **Code Download**.

You can also download the code files by clicking on the **Code Files** button on the book's webpage at the Packt Publishing website. This page can be accessed by entering the book's name in the **Search** box. Please note that you need to be logged in to your Packt account.

Once the file is downloaded, please make sure that you unzip or extract the folder using the latest version of:

- WinRAR / 7-Zip for Windows
- Zipeg / iZip / UnRarX for Mac
- 7-Zip / PeaZip for Linux

The code bundle for the book is also hosted on GitHub at `https://github.com/PacktPublishing/Building-Smart-Homes-with-Raspberry-Pi-Zero`. We also have other code bundles from our rich catalog of books and videos available at `https://github.com/PacktPublishing/`. Check them out!

Errata

Although we have taken every care to ensure the accuracy of our content, mistakes do happen. If you find a mistake in one of our books—maybe a mistake in the text or the code—we would be grateful if you could report this to us. By doing so, you can save other readers from frustration and help us improve subsequent versions of this book. If you find any errata, please report them by visiting `http://www.packtpub.com/submit-errata`, selecting your book, clicking on the **Errata Submission Form** link, and entering the details of your errata. Once your errata are verified, your submission will be accepted and the errata will be uploaded to our website or added to any list of existing errata under the Errata section of that title.

To view the previously submitted errata, go to `https://www.packtpub.com/books/content/support` and enter the name of the book in the search field. The required information will appear under the **Errata** section.

Piracy

Piracy of copyrighted material on the Internet is an ongoing problem across all media. At Packt, we take the protection of our copyright and licenses very seriously. If you come across any illegal copies of our works in any form on the Internet, please provide us with the location address or website name immediately so that we can pursue a remedy.

lease contact us at copyright@packtpub.com with a link to the suspected pirated material.

We appreciate your help in protecting our authors and our ability to bring you valuable content.

Questions

If you have a problem with any aspect of this book, you can contact us at questions@packtpub.com, and we will do our best to address the problem.

1
Configuring Your Raspberry Pi Zero Board

In the first chapter of this book, we are going to go through all the steps that are required to configure your Raspberry Pi Zero board so you can use it for all the projects we will see in this book.

First we will look at the list of components that are required to use the board. Then, we will install the Raspbian distribution, which will be the operating system we will use throughout this book, on the board. Finally, we'll see how to configure the board for remote, and how to install the Node.js framework that we will use in nearly all the projects of the book. Let's start!

Introducing the Raspberry Pi Zero board

The Raspberry Pi Zero is a board that was introduced in 2015, and the goal was to make a low-cost ($5), small-format board with most of the functionalities of the original Raspberry Pi board.

The following is an image of the Zero board:

In the center of the board, you will find the same **System-on-a-Chip (SoC)** as the original Raspberry Pi board, with a 1-GHz single-core processor, 512 MB of RAM, and a graphical processing unit.

The board has several inputs and outputs, like the 40-pin GPIO connector that we will use through this whole book to connect the board to sensors and other components.

There are also two USB ports (one for power, one for communication), one mini-HDMI port, and one SD card slot to put the operating system and other files in.

The power consumption of the board was also reduced compared to the first board, going from 1.5W to 0.8W.

Required components for the Zero board

Even if the Raspberry Pi Zero board has a very small form factor, it actually can't be used alone, at least for the configuration step. Therefore, we are going to need a lot of additional components for all the projects of this book, and this is what I wanted to go through in this section.

The first thing you will need for your Pi Zero board is a micro-USB to USB converter, so you can plug regular USB devices into your board. This is an image of the cable I used for my Pi board:

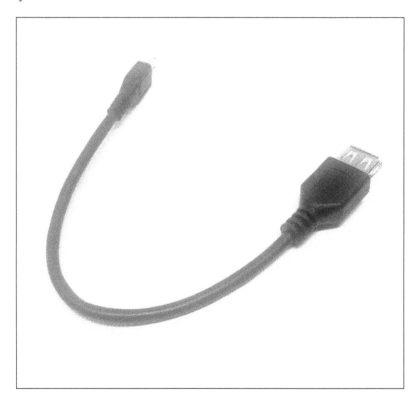

Then, you will need some way to connect your Raspberry Pi Zero board to a computer screen. To do so, you will need a mini-HDMI to HDMI adapter:

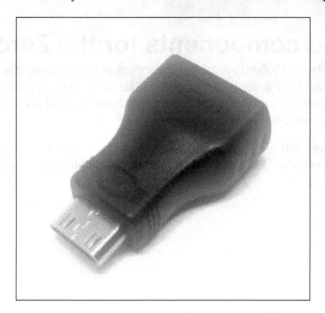

In order to connect more than one device to the board, you will also need a regular USB hub:

Later in this chapter, we are going to look at how to use the Raspberry Pi board remotely from your computer, so you don't need to always have it connected to an external screen. However, to begin with, you will need a keyboard and mouse to use it:

 I recommend using a keyboard with a small track pad as well.

The Raspberry Pi Zero board doesn't come with onboard storage. Therefore, you will need to use a micro SD card to store the operating system. I recommend using at least an 8-GB SD card:

At some point, we are going to connect the Raspberry Pi Zero board to the Internet. We'll also need to connect it to your local network, so you can access it remotely. To do so, I recommend using a simple Wi-Fi USB dongle:

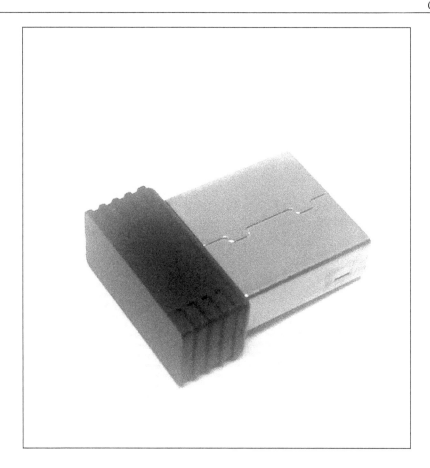

Of course, you will need other components to use the Raspberry Pi Zero board that I haven't included here as usually, they are already on everyone's desk. For example, you will need a screen with an HDMI input to use the board. You will also need a micro-USB power supply, for example, the one you are using to charge your phone. The Raspberry Pi foundation says it should work with a power supply that can deliver at least 1A, but a 2A power supply is recommended.

Assembling the different components

Let's now look at how to assemble the required components so we can get started with your Raspberry Pi:

1. First, insert the micro-USB to USB adapter cable into one of the USB ports of the Pi (not the PWR one), and also plug the mini-HDMI cable to the Pi.

2. Then, connect all your required USB devices (for example, the Wi-Fi dongle) to the USB hub, and connect the USB to the Pi. Also, connect the Pi to an external screen using an HDMI cable.

3. This is how it should look when you are done, not showing the connections to the screen or the hub:

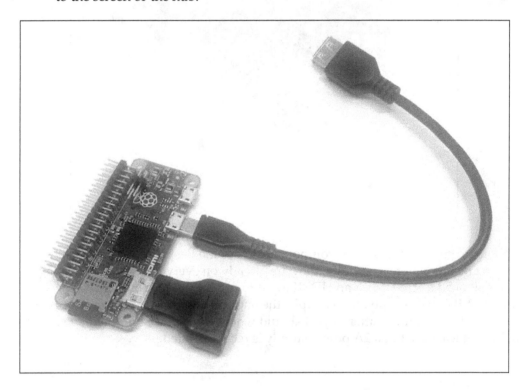

4. At this point, don't connect it to the power yet—we first need to install Raspbian (the operating system) on the SD card.

Installing Raspbian

There are many operating systems that are available for the Raspberry Pi board, most of which are based on Linux. However, the one that is usually recommended is Raspbian, which is an operating system based on Debian that was specifically made for Raspberry Pi.

In order to install the Raspbian operating system on your Pi, the first step is to download the latest Raspbian image from the official Raspberry Pi website:

```
https://www.raspberrypi.org/downloads/raspbian/
```

Next, insert the micro SD card into your computer using an adapter (an adapter is usually given with the SD card). To actually configure the SD card, it's best to refer to the official installation guides. If you use Windows, please refer to the following URL:

```
https://www.raspberrypi.org/documentation/installation/installing-
images/windows.md
```

If you are using OS X, please refer to the following:

```
https://www.raspberrypi.org/documentation/installation/installing-
images/mac.md
```

Finally, if you are using Linux, you can refer to the following:

```
https://www.raspberrypi.org/documentation/installation/installing-
images/linux.md
```

Now, once you have Raspbian installed on your SD card, insert it into Raspberry Pi and connect the Raspberry Pi board to the power source via the micro-USB port.

Then, after a while, you should see the desktop of your freshly installed Raspbian operating system:

If you can see this screen, congratulations; you now have a fully functional Raspberry Pi Zero board. Throughout the rest of this chapter, we are going to see how to configure the board so it can be accessed remotely, and how to install the Node.js framework on it.

Configuring the board for remote access

At the end of this chapter, you want to be able to access the board from your own computer, without having it connected to an external screen.

The first step is to connect the Raspberry Pi board to your local Wi-Fi network. If you connected a Wi-Fi dongle to the Pi, you should see the Wi-Fi icon at the top of your Pi desktop. Click on it, and select your Wi-Fi network:

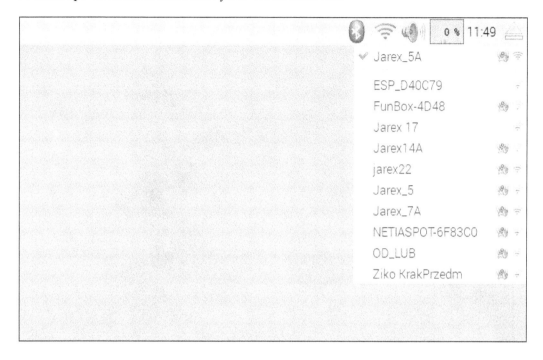

You will then be asked to enter the password for your network, and a few seconds later, you should be connected to your local Wi-Fi network and to the Internet.

Next, we need to enter the Raspberry Pi configuration panel so we can set some essential settings. You can find the **Raspberry Pi Configuration** tool inside the main **Menu**:

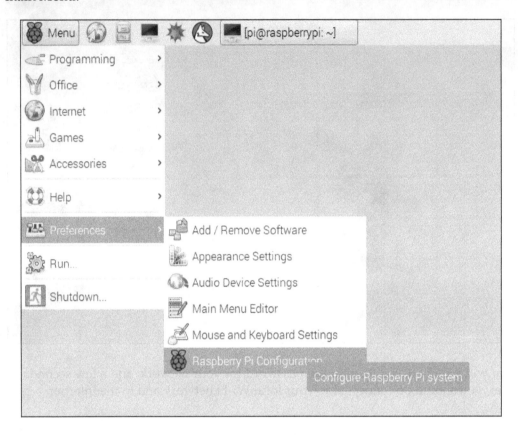

Inside the configuration tool, first press on **Expand Filesystem**:

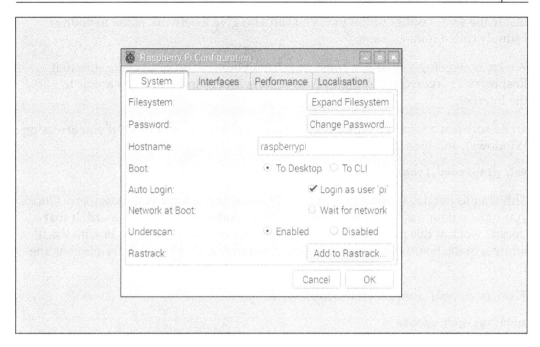

This will make sure that the Pi now has access to all the space available on the SD card. Also, click on the **Interfaces** tab and check that **SSL** is checked:

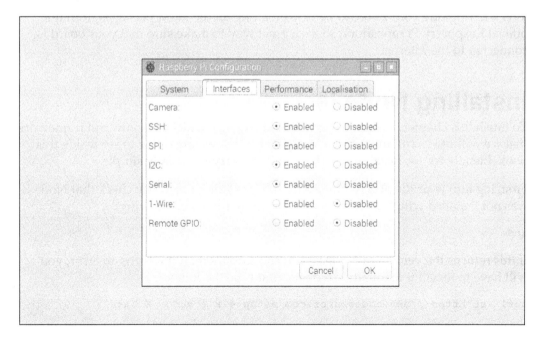

Inside the same configuration tool, you can also give a network name to your Pi. I simply called mine `pizero`.

We are now going to perform some tests from your computer to make sure that Raspberry Pi is correctly configured for remote access and that it has access to the Internet.

To do so, open a terminal window on your computer (or use PuTTY if you are using Windows), and type the following:

```
ssh pi@pizero.local
```

This should initiate a connection to your Pi board and ask for your password. Once you type in your password, you should now be connected to the Pi board. If that doesn't work at this point, try replacing the name of your Raspberry Pi with the IP address of the board (you can get this by typing `ifconfig` inside a Terminal on the Pi itself).

Then, from your computer, type the following:

```
sudo apt-get update
```

Then type the following command:

```
sudo apt-get upgrade
```

This will upgrade your Pi board by downloading all the latest packages from the official Raspberry Pi repository, so it's a great way to make sure that your board is connected to the Internet.

Installing Node.js

To finish this chapter, we are going to install Node.js, which is a powerful framework that we will use to run most of the applications that we are going to see inside this book. Luckily for us, installing Node.js on Raspberry Pi is really simple.

First, log into your Raspberry Pi via SSH. We are going to quickly check that Node.js was not installed with the Linux image. To do so, type the following:

```
node -v
```

If this returns the version of Node.js, you can stop there. If it returns an error, you will have to install it manually. To do so, first type the following:

```
curl -sL https://deb.nodesource.com/setup_4.x | sudo -E bash -
```

After that, type the following command, which will install Node.js:

```
sudo apt-get install -y nodejs
```

Finally, install some additional tools with the following:

```
sudo apt-get install -y build-essential
```

You can now test that Node.js is correctly installed by typing the following:

```
node -v
```

Congratulations, you now have a fully configured Raspberry Pi Zero! In the next chapter, we are going to build your first application using the Zero board and learn how to measure data from sensors.

Summary

In this first chapter of this book, we saw how to configure our Raspberry Pi Zero board so we can use it in later chapters of this book. We saw what components were needed for the Pi, and how to install Raspbian so we can run software on our board. Finally, we installed Node.js, which we will use in the whole book to run home automation projects on our Pi.

In the following chapter, we are going to dive into the core topic of the book by learning how to measure data from a sensor.

2
Measure Data Using Your Raspberry Pi Zero Board

In the first chapter of this book, we worked on setting up your Raspberry Pi board so you can use it in your projects and realize all the projects you'll find in this book.

In this chapter, we are going to make our first project using the Zero board: measuring data using your board. We are going to learn how to connect a very simple temperature and humidity digital sensor to your Pi, and how to write software to read data from it.

From there, we'll look at some very basic applications using this sensor that can be really useful inside a smart home: how to log data on the Pi itself, how to access the measurements remotely, and finally, how to display past data on a nice plot.

Hardware and software requirements

We have already discussed most of the requirements for this project in the first chapter of this book. Here, you will simply need an additional component: a DHT11 sensor (`https://www.adafruit.com/products/386`). The following image shows the sensor:

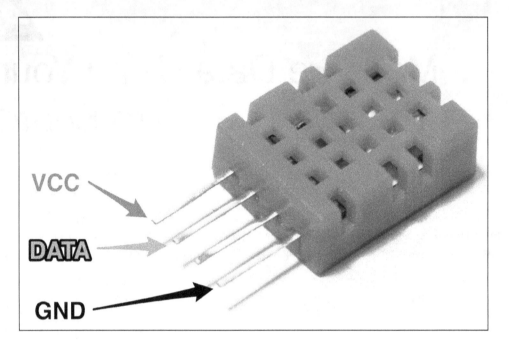

You can of course use other similar sensors, for example the DHT22, which is more precise. To use a DHT22, you will only need to change one thing inside the code we'll see later.

You will also need a 4.7k Ohm resistor to make the sensor work, as well as jumper wires and a breadboard.

Hardware configuration

Let's now look at how to configure the hardware for this project; basically, how to connect the sensor to the Pi Zero board.

The following figure is a schematic to help you out:

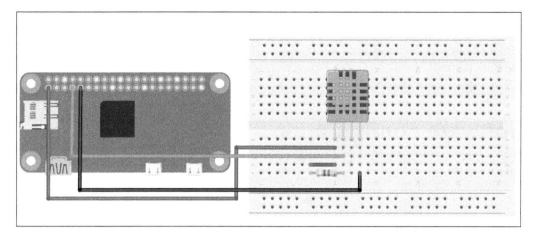

As it's the first project we are actually building using the Raspberry Pi Zero, there is something important I wanted to point out here. To connect the board to components like this sensor here, we have two options. You can either use jumper wires directly (as shown on the schematic), or use a cobbler kit to connect all the pins of the Pi to the breadboard, as shown in the following image:

This is up to you, and to be clear, I'll always show only the individual wires on the schematics, but use a cobbler kit to actually build the projects.

Here, you simply need to place the DHT11 on the breadboard, and then connect the resistor between the VCC and the data pins. Then, connect the VCC to the 3.3V pin of the Raspberry Pi, GND to GND, and finally, connect the data pin of the sensor to pin 4 of the Raspberry Pi board.

Software configuration

Now we are going to install additional software on your Pi to make sure we can read data from the sensor.

Following the instructions from *Chapter 1, Configuring Your Raspberry Pi Zero Board* log into your Pi via SSH, or just use it with an external screen with mouse and keyboard.

1. Inside a terminal, type the following:

   ```
   wget http://www.airspayce.com/mikem/bcm2835/bcm2835-1.50.tar.gz
   ```

2. Wait for the download to complete and then type the following:

   ```
   tar zxvf bcm2835-1.50.tar.gz
   ```

3. Next, type the following:

   ```
   cd bcm2835-1.50
   ```

4. Now, configure the software you just downloaded with the following:

   ```
   ./configure
   ```

5. Build this software with the following:

   ```
   make
   ```

6. Now, verify that everything is okay with the following:

   ```
   sudo make check
   ```

7. If there are no errors, you can then install the software on your Pi with the following:

   ```
   sudo make install
   ```

After that last step, you can now move on to the projects of this chapter!

Reading data from the sensor

As the first project of this chapter, we are simply going to see how to read data from the sensor. As for all the projects in this book, we'll use Node.js, which is a great framework for building projects on your Raspberry Pi Zero.

I will now go through the main parts of this first piece of code. It starts by including the DHT sensor module for Node.js:

```
var sensorLib = require('node-dht-sensor');
```

Then, we create an object to read data from the sensor and initialize it when we start the software:

```
var sensor = {
    initialize: function () {
        return sensorLib.initialize(11, 4);
    },
    read: function () {
        var readout = sensorLib.read();
        console.log('Temperature: ' + readout.temperature.toFixed(2) +
'C, ' +
            'humidity: ' + readout.humidity.toFixed(2) + '%');
        setTimeout(function () {
            sensor.read();
        }, 2000);
    }
};

if (sensor.initialize()) {
    sensor.read();
} else {
    console.warn('Failed to initialize sensor');
}
```

You can now either copy the code inside a file called sensor_test.js, or just get the complete code from the GitHub repository for this project:

```
https://github.com/openhomeautomation/smart-homes-pi-zero
```

Next, use Terminal to navigate to the folder where the files are and type the following:

```
npm install node-dht-sensor
```

This will install the module to read data from the sensor; it can take a while, so be patient. In case it doesn't work, try using sudo in front of the command. Next, actually start the software with the following:

```
sudo node sensor_test.js
```

This should print the readings of the sensor at regular intervals inside the terminal:

```
pi@raspberrypi:~/Work/smart-homes-pi-zero/01 $ sudo node sensor_test.js
Temperature: 26.00C, humidity: 33.00%
Temperature: 26.00C, humidity: 33.00%
Temperature: 26.00C, humidity: 33.00%
Temperature: 26.00C, humidity: 33.00%
Temperature: 26.00C, humidity: 33.00%
Temperature: 26.00C, humidity: 33.00%
Temperature: 26.00C, humidity: 33.00%
Temperature: 26.00C, humidity: 33.00%
Temperature: 26.00C, humidity: 33.00%
Temperature: 26.00C, humidity: 33.00%
Temperature: 26.00C, humidity: 33.00%
Temperature: 26.00C, humidity: 33.00%
Temperature: 26.00C, humidity: 33.00%
Temperature: 26.00C, humidity: 33.00%
Temperature: 26.00C, humidity: 33.00%
Temperature: 26.00C, humidity: 33.00%
Temperature: 26.00C, humidity: 33.00%
```

Congratulations, you can now read data from a digital sensor using your Pi Zero board! This is the first step to building sensors for your smart home.

Storing sensor data

Displaying the current measurements from the sensor is nice, but what is even better is to actually store that data inside a database. In this section, we are going to see how easy it is to do this with Node.js.

As a database, we'll simply use NeDB here, which is a really simple database for Node.js that is completely stored in memory, but you can also save the entire database in a file.

The code is actually very similar to what we saw in the previous section. However, here, we'll first import the database module, and then insert data inside the database when a measurement is done:

```
var Datastore = require('nedb')
  , db = new Datastore({ filename: 'path/to/datafile', autoload: true
});
sdfsd
var readout = sensorLib.read();

// Log
var data = {
    humidity: readout.humidity.toFixed(2),
    temperature: readout.temperature.toFixed(2),
    date: new Date()
};
db.insert(data, function (err, newDoc) {
    console.log(newDoc);
});

// Repeat
setTimeout(function () {
    sensor.read();
}, 2000);
```

You can of course find all the code inside the GitHub repository of the book. Again, navigate to the folder where the files are located and type the following:

```
npm install nedb --save
```

This will install the NeDB module for Node.js.

Then, start the recording with the following:

```
sudo node sensor_record.js
```

You should see the measurements being recorded at regular intervals:

```
pi@raspberrypi:~/Work/smart-homes-pi-zero/01 $ sudo node sensor_record.js
{ humidity: '34.00',
  temperature: '24.00',
  date: Sun May 22 2016 09:18:26 GMT+0000 (UTC),
  _id: 'c7zhxmlJNZgknPPb' }
{ humidity: '34.00',
  temperature: '24.00',
  date: Sun May 22 2016 09:18:29 GMT+0000 (UTC),
  _id: '0RpgicQscdBSmeGM' }
{ humidity: '34.00',
  temperature: '24.00',
  date: Sun May 22 2016 09:18:32 GMT+0000 (UTC),
  _id: 'K2gpjaKZDYggDkI5' }
{ humidity: '34.00',
  temperature: '24.00',
  date: Sun May 22 2016 09:18:35 GMT+0000 (UTC),
  _id: 'CCG4kxgxQjIKeonL' }
```

Now, we didn't learn how to actually retrieve those measurements, but that's something we will see later in this chapter. In the meantime, if you want more information about how to retrieve documents, you can look at the official page on GitHub:

```
https://github.com/louischatriot/nedb
```

Accessing the data remotely

In the previous projects of this chapter, we learned how to measure and store data on your Pi. However, in a smart home, the best is to be able to access data remotely, for example, from your smartphone or computer. We will see many similar examples in later chapters of this book, but in this chapter, I just wanted to give you a glimpse of what is possible.

The module we are going to use here is Express, a server framework that is really easy to use with Node.js. Express works by defining routes, which is what will be served to the client if a request is made on a specific URL.

First, we'll import Express and define a main route that will send back the temperature and humidity measurements:

```
var express = require('express');
var app = express();
```

```
app.get('/', function (req, res) {

    var readout = sensor.read();
    answer = 'Temperature: ' + readout.temperature.toFixed(2);
    answer += ' Humidity: ' + readout.humidity.toFixed(2);
    res.send(answer);

});
```

Finally, we also need to start the application, and once that's done we print a message in the console:

```
app.listen(3000, function () {
    console.log('Raspberry Pi Zero app listening on port 3000!');
});
```

It's now time to test our little web server! Once you have grabbed the file from the book's GitHub repository, navigate to the folder where the files are and type the following:

npm install express

Then, launch the app with the following:

sudo node sensor_express.js

You should get the confirmation inside the console. Now, using your computer or a smartphone, navigate to the URL of your Pi, not forgetting to add port 3000: http://192.168.0.105:3000/.

If you don't know the IP address of your Pi, you can simply type ifconfig while logged in.

The page should display the last measurement made by the Pi:

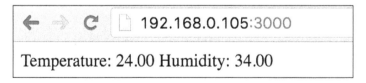

Of course, this can definitely be improved with a much nicer interface to display the measurements. However, in this section the goal was really to show you how to display those measurements from a device other than the Pi.

Plotting the stored data

In the final project of this chapter, we are going to learn how to plot the data that was measured by the Raspberry Pi Zero board. We are actually going to combine what we did in the other projects of this chapter and add the plotting part on top of that.

As the code is quite similar to what we have already seen, I will only highlight the main changes here. First, we need to define a route for the data:

```
app.get('/data', function (req, res) {

  db.find({}, function (err, docs) {

    res.json(docs);

  });

});
```

This will make sure that, when it is queried on this route, the server will return all the measurements stored so far inside the database.

Then, to display the plot of all the measurements, we are going to use a JavaScript called **HighCharts**. You can find more information about HighCharts here:

http://www.highcharts.com/

We'll include it inside an HTML file that we will place inside a folder called public, so our app can access it. This file will basically import all the JavaScript libraries that we need, another script, which we'll see in a moment, and a container for the plot:

```
<!DOCTYPE html>
<html>

<head>
  <script src="https://code.jquery.com/jquery-2.2.4.min.js"></script>
  <script src="https://code.highcharts.com/highcharts.js"></script>
  <script src="https://code.highcharts.com/modules/exporting.js"></
script>
  <script src="js/script.js"></script>
</head>

<body>
```

```
<div id="container" style="min-width: 310px; height: 400px; margin: 0
auto"></div>

</body>
</html>
```

Now, we also need to make the link between the HTML page and our application. This will be done in a script file called `script.js`. This is the content of this file:

```
var dates = [];
var temperature = [];
var humidity = [];

console.log(measurements);

for (i = 0; i < measurements.length; i++) {

    dates.push(measurements[i].date);
    temperature.push(parseFloat(measurements[i].temperature));
    humidity.push(parseFloat(measurements[i].humidity));

}
$('#container').highcharts({
        title: {
            text: 'Temperature & Humidity Data',
            x: -20 //center
        },
        xAxis: {
            categories: dates
        },
        yAxis: {
            title: {
                text: 'Temperature (°C)'
            },
            plotLines: [{
                value: 0,
                width: 1,
                color: '#808080'
            }]
        },
        tooltip: {
            valueSuffix: '°C'
        },
```

```
        legend: {
              layout: 'vertical',
              align: 'right',
              verticalAlign: 'middle',
              borderWidth: 0
        },
        series: [{
              name: 'Temperature',
              data: temperature
        },
        {
              name: 'Humidity',
              data: humidity
        }]
    });
```

Basically, this file will query the most recent data from the application, format it for HighCharts, and then actually plot the data.

You can now grab all the files from the book's GitHub repository and start the application with the following:

```
sudo node sensor_plot.js
```

The first thing you can do is test the data route by going to the IP address of your Pi, followed by port 3000 and the data route:

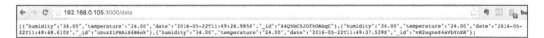

As you can see, all the measurements made so far are extracted from the database and returned by the server.

You can again go to the main route, and you should see the same data on a nice plot:

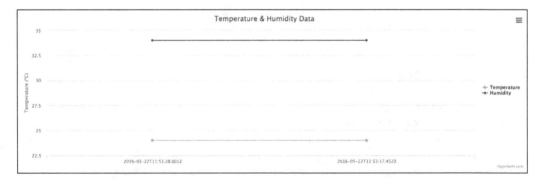

If you wait a bit more, you'll see a nice graph with all the data recorded so far by the application running on your Raspberry Pi board:

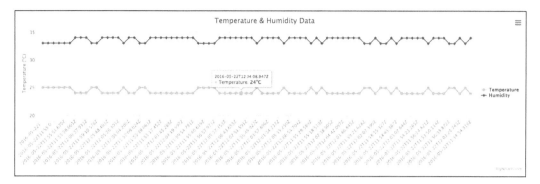

Of course, you can adjust some settings, for example, the delay between two measurements, inside the code for your own projects.

Summary

In this chapter, we saw how to perform a basic task with the Raspberry Pi Zero board: measuring data. We saw how to measure data from a digital sensor, and then store this data, access it remotely, and finally even plot the data on a graph.

You can of course already use what you learned in this project and adapt it to your own projects. You can, for example, have the project measure from more sensors at the same time, for example, from a barometric pressure sensor or from a light-level sensor.

In the next chapter, we are going to apply what we have learned in this chapter to build another project with the Pi Zero: building your own home thermostat.

3
Building a Smart Home Thermostat

In the previous chapter, we learned how to read data from a sensor and log this data on the Raspberry Pi Zero board. In this chapter, we are going to use that knowledge to build a very useful home-automation component: a smart thermostat.

We are going to see how to use the Raspberry Pi Zero and a few other components to regulate the temperature in a room of your home using an electrical heater. We'll see how to connect all the different components, and also how to create a nice interface that you will be able to use to control your thermostat. Let's start!

Hardware and software requirements

As always, we are going to start with a list of required hardware and software components for the project.

Except for the Raspberry Pi Zero, the most important component for this project will be the PowerSwitch Tail Kit. This component allows your Pi to control electrical appliances such as lamps, heaters, and other appliances that use mains electricity to function.

Then, we will use the same DHT11 sensor we used in the previous chapter to measure the temperature in the room.

Finally, you will need the usual breadboard and jumper wires.

This is the list of components you will need for this project, not including the Raspberry Pi Zero:

- PowerSwitch Tail Kit (`https://www.adafruit.com/products/268`)
- DHT11 sensor + 4.7k Ohm resistor (`https://www.adafruit.com/products/386`)
- Breadboard (`https://www.adafruit.com/products/64`)
- Jumper wires (`https://www.adafruit.com/products/1957`)

Of course, for the project to make sense, you will need to have an electrical heater that you can control. As I don't want you to cut any wires from an existing heater for this project, I recommend trying it first, using a portable electrical heater that you can find in any shop, like this one:

The PowerSwitch Tail component supports up to 1800W of power, so you can choose your heater accordingly, up to this limit.

On the software side, you don't need anything else compared to the previous chapter.

Hardware configuration

Let's now see how to configure the hardware for this project; basically, how to connect the PowerSwitch Tail and the sensor to the Raspberry Pi Zero board.

The following is a schematic to help you out:

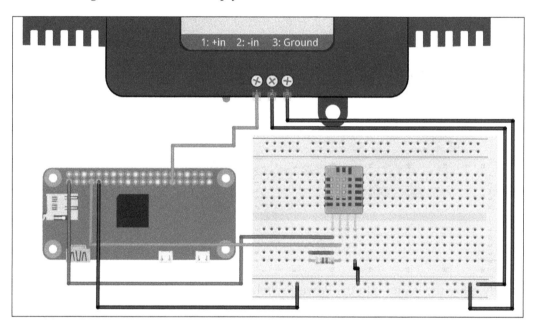

When done, the sensor will be connected to GPIO pin 4, and the heater (via the PowerSwitch Tail) to pin 29.

Here, you simply need to place the DHT11 on the breadboard, and then connect the resistor between the VCC and the data pins. Then, connect the VCC to the 3.3V pin of the Raspberry Pi, GND to GND, and finally, connect the data pin of the sensor to pin 4 of the Raspberry Pi board.

For the PowerSwitch Tail, simply connect the Vin+ pin of the component to pin 29 of the Raspberry Pi, and then the remaining two pins of the PowerSwitch Tail to the ground.

Here again, I chose to represent the individual wires on the schematic for the purpose of clarity, but I used a cobbler cable on the project itself. Here is the final result:

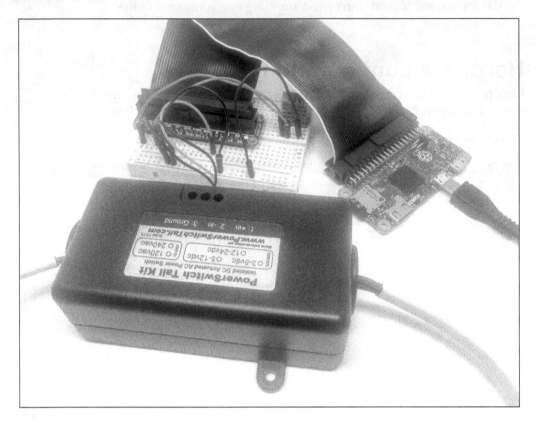

Finally, plug the heater into the PowerSwitch Tail and connect the PowerSwitch Tail to the mains electricity.

Testing individual components

As the first project of this chapter, we are simply going to check that each individual component (the sensor and the PowerSwitch tail) are working correctly.

I will now go through the main parts of this first piece of code. It starts by including the DHT sensor module for Node.js:

```
var sensorLib = require('node-dht-sensor');
```

Then, we create an object to read data from the sensor, and also initialize it when we start the software:

```
var sensor = {
    initialize: function () {
        return sensorLib.initialize(11, 4);
    },
    read: function () {
        var readout = sensorLib.read();
        console.log('Temperature: ' + readout.temperature.toFixed(2) +
'C, ' +
            'humidity: ' + readout.humidity.toFixed(2) + '%');
        setTimeout(function () {
            sensor.read();
        }, 2000);
    }
};

if (sensor.initialize()) {
    sensor.read();
} else {
    console.warn('Failed to initialize sensor');
}
```

You can now either copy the code inside a file called `sensor_test.js`, or just get the complete code from the GitHub repository of the project:

`https://github.com/openhomeautomation/smart-homes-pi-zero`

Next, use the terminal to navigate to the folder where the files are, and type the following:

`npm install node-dht-sensor`

This will install the module to read data from the sensor; it can take a while, so be patient. In case it doesn't work, try using `sudo` in front of the command. Next, actually start the software with the following command:

`sudo node sensor_test.js`

This should print the readings of the sensor at regular intervals inside the terminal:

```
pi@raspberrypi:~/Work/smart-homes-pi-zero/03 $ ls
lamp_test.js  node_modules  public  sensor_server.js  sensor_test.js
pi@raspberrypi:~/Work/smart-homes-pi-zero/03 $ sudo node sensor_test.js
Temperature: 27.00C, humidity: 32.00%
Temperature: 27.00C, humidity: 32.00%
Temperature: 27.00C, humidity: 32.00%
^Cpi@raspberrypi:~/Work/smart-homes-pi-zero/03 $
```

We are now going to see how to test if the PowerSwitch Tail is working and wired correctly, and how to control it remotely. For that, we are going to use the aREST module for the Raspberry Pi, which will give us an easy way to control the outputs of the board.

Here is the complete code for this part:

```
// Start
var express = require('express');
var app = express();
var piREST = require('pi-arest')(app);

piREST.set_id('34f5eQ');
piREST.set_name('my_rpi_zero');

var server = app.listen(80, function() {
    console.log('Listening on port %d', server.address().port);
});
```

The code is pretty simple, and we are going to try it right now. First, you need to install the required modules by typing the following in a terminal (where the files of the project are located):

```
sudo npm install express pi-arest
```

Then, simply start the application with the following command:

```
sudo node heaer_test.js
```

You should now see the confirmation in your console.

Now, let's go ahead and try to control the heater, for example, to turn it on. You need to make sure that it is actually turned on if there is any mechanical switch on the heater itself.

Then, go to your favorite web browser, and type the following:

```
http://raspberrypi.local/digital/29/1
```

You should see that the heater turns on instantly, and you should also have a confirmation inside your web browser. Then, type the following command to turn it off again:

```
http://raspberrypi.local/digital/29/0
```

If that works, you can now control your heater from your Raspberry Pi Zero! You can now move to the next section, in which we are going to code the thermostat.

Building the thermostat

We are now going to see how to build the code for the thermostat, which will run on your Raspberry Pi Zero board. As the code is quite long, I will only highlight the most important parts here, but you can of course find the complete code inside this book's GitHub repository.

Start by importing the required modules:

```
var sensorLib = require('node-dht-sensor');
var express = require('express');
```

Then, we create an Express app, which will allow us to easily structure our application:

```
var app = express();
```

Next, we define some variables that are important for our thermostat:

```
var targetTemperature = 25;
var threshold = 1;
var heaterPin = 29;
```

The threshold is here so the thermostat doesn't constantly switch between the on and off states when it is near the target temperature. A lower threshold means that you will have a temperature closer to what you want, but also that the heater will switch more frequently.

After that, we are going to define the routes that will structure our application. The first one is a route to get the thermostat's current target temperature:

```
app.get('/get', function (req, res) {

  answer = {
    targetTemperature: targetTemperature
  };
  res.json(answer);

});
```

We will also define another route to set this target temperature, which will be called by the interface we will code in a moment:

```
app.get('/set', function (req, res) {

  // Set
  targetTemperature = req.query.targetTemperature;

  // Answer
  answer = {
    targetTemperature: targetTemperature
  };
  res.json(answer);

});
```

Finally, we also need a route to get the current value of the temperature by performing a measurement on the sensor:

```
app.get('/temperature', function (req, res) {

  answer = {
    temperature: sensor.read().temperature.toFixed(2)
  };
  res.json(answer);

});
```

Now, we also need to integrate all the code that we will use to control the heater from the Raspberry Pi. We saw this before, when we tested the PowerSwitch Tail:

```
var piREST = require('pi-arest')(app);
piREST.set_id('34f5eQ');
piREST.set_name('my_rpi_zero');
sd
app.listen(3000, function () {
  console.log('Raspberry Pi Zero thermostat started!');
});
```

We still need to write the code for the core of the thermostat function. Indeed, we want the Pi Zero to regulate the temperature in your home, whether you are currently using the interface or not. This is done with the following piece of code:

```
setInterval(function () {

  // Check temperature
```

```
temperature = parseFloat(sensor.read().temperature);
console.log('Current temperature:' + temperature);
console.log('Target temperature: ' + parseFloat(targetTemperature));

// Too high?
if (temperature > parseFloat(targetTemperature) + 1) {
  console.log('Deactivating heater');
  piREST.digitalWrite(heaterPin, 0);
}

// Too low?
if (temperature < parseFloat(targetTemperature) - 1) {
  console.log('Activating heater');
  piREST.digitalWrite(heaterPin, 1);
}

}, 10 * 1000);
```

Basically, we check every 10 seconds and compare the current temperature to the target temperature defined inside the thermostat. If it's too low, for example, we activate the heater.

Finally, we also define the function to read data from the temperature sensor:

```
var sensor = {
    initialize: function () {
        return sensorLib.initialize(11, 4);
    },
    read: function () {

        // Read
        var readout = sensorLib.read();
        return readout;
    }
};

if (sensor.initialize()) {
    sensor.read();
} else {
    console.warn('Failed to initialize sensor');
}
```

It's now time to test the thermostat! Make sure to grab all the code from this book's GitHub repository, navigate to the folder for this chapter, and type the following:

```
npm install node-dht-sensor
```

Then type the following command:

```
sudo npm install express pi-arest
```

You can then start the project with the following command:

```
sudo node thermostat_server.js
```

You should immediately see a message similar to the following on the console:

```
Raspberry Pi Zero thermostat started!
Current temperature:27
Target temperature: 25
Deactivating heater
Current temperature:27
Target temperature: 25
Deactivating heater
```

You can now test all the routes we defined earlier:

- For example, get the temperature with the following: `http://raspberrypi.local:3000/temperature`
- You can also get the current value of the thermostat with the following: `http://raspberrypi.local:3000/get`
- Finally, you can set the target of the thermostat using the following: `http://raspberrypi.local:3000/set?targetTemperature=20`For example, set it to a high value: `http://raspberrypi.local:3000/set?targetTemperature=30`

You should quickly see the thermostat reacting to this new target by activating the heater:

```
Current temperature:27
Target temperature: 30
Activating heater
Current temperature:27
Target temperature: 30
Activating heater
```

This is great, but every modern thermostat has some kind of interface where you can set the temperature of the thermostat. This is exactly what we are going to do in the final part of this chapter.

Controlling the thermostat remotely

We are now going to take the exact same project we defined earlier, but add a graphical interface on top of it. Inside the JavaScript file we saw previously, you just need to add one line, as follows:

```
app.use(express.static('public'));
```

Now we are going to code two files: one HTML file with the interface, and another file containing scripts that will make the link between the interface and the server. Let's start with the HTML file:

```
<head>
  <script src="https://code.jquery.com/jquery-2.2.4.min.js"></script>
  <link rel="stylesheet" href="https://maxcdn.bootstrapcdn.com/
bootstrap/3.3.6/css/bootstrap.min.css">
  <script src="https://maxcdn.bootstrapcdn.com/bootstrap/3.3.6/js/
bootstrap.min.js"></script>
  <script src="js/script.js"></script>
  <link rel="stylesheet" href="css/style.css">
  <meta name="viewport" content="width=device-width, initial-scale=1">
</head>
```

As you can see, inside the `<head>` tag of this file, we basically import components such as jQuery and Bootstrap, and also a file called `script.js`. It is in this file that we will place all the JavaScript functions later on.

Then, we define a first row for the current value of the temperature in the room:

```
<div class='row'>

    <div class='col-md-4'></div>
    <div class='col-md-4 text-center'>
      Current temperature value:
      <span id='temperature'></span> C</div>
    <div class='col-md-4'></div>

</div>
```

We then do the same with the target temperature:

```
<div class='row'>

    <div class='col-md-4'></div>
    <div class='col-md-4 text-center'>
      Current thermostat value:
      <span id='thermostat'></span> C</div>
    <div class='col-md-4'></div>

  </div>
```

After that, we create a text input that will be used to input a new target temperature:

```
<div class='row'>

    <div class='col-md-4'></div>
    <div class='col-md-4'>
      <input type="text" class="form-control" id="thermostatValue"></
div>
    <div class='col-md-4'></div>

  </div>
```

Finally, we create a button so the user can validate this new temperature:

```
<div class='row'>

    <div class='col-md-4'></div>
    <div class='col-md-4'>
      <button id='set-thermostat' class='btn btn-block btn-primary'>
        Set Thermostat</button></div>
    <div class='col-md-4'></div>

  </div>
```

Now, let's have a look at the file called `script.js`, which will make the link between the interface and the server. First, we refresh the target temperature and the current temperature inside the interface:

```
$.get('/get', function(data) {

    $('#thermostat').html(data.targetTemperature);

});

$.get('/temperature', function(data) {

    $('#temperature').html(data.temperature);

});
```

Next, we set the new value of the target temperature whenever we click on the button:

```
$( "#set-thermostat" ).click(function() {

    // Get value
    var newThermostatValue = $('#thermostatValue').val();

    // Set new value
    $.get('/set?targetTemperature=' + newThermostatValue,
function(data) {
        $('#thermostat').html(data.targetTemperature);
    });

});
```

It's now finally time to test the project! Simply start it with the following code:

```
sudo node thermostat_interface.js
```

Now, go to your web browser, and type the following:

```
http://raspberrypi.local/interface.html
```

You should immediately see the following interface:

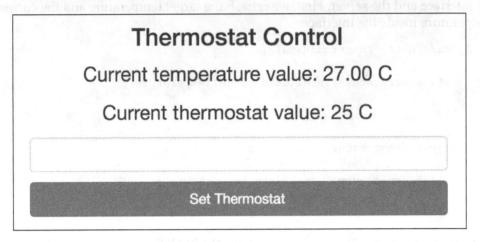

You can now try this interface, for example, by typing the value of a new target for the thermostat:

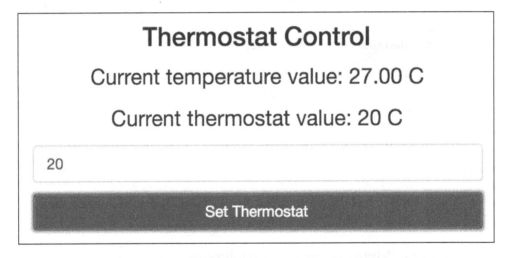

If you want to see if it is functioning correctly, simply type a value that is high compared to the current value of the temperature:

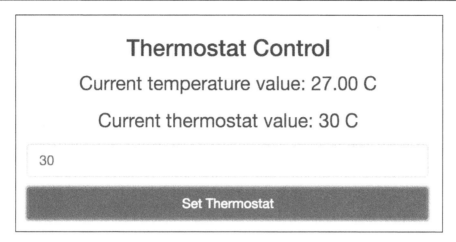

You should immediately see that the heater is turning on and that the temperature starts to rise after a while:

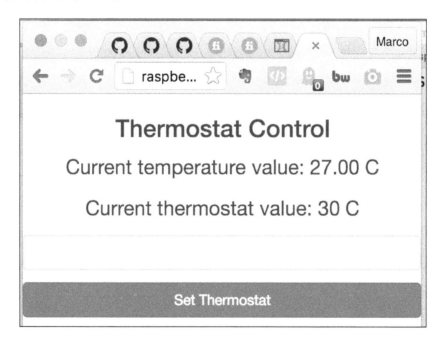

Congratulations, you just built your own thermostat based on the Raspberry Pi Zero!

Summary

In this chapter, we saw how to build a simple thermostat using the Raspberry Pi Zero board. We were able to make a thermostat that can generate its own interface and we are able to set the target temperature of the thermostat via this interface.

You can of course already use what you learned in this project, and adapt it to a heater you have in your home. I recommend you first try it on a portable heater, and then to install it on an actual heater on your wall, if you are feeling confident with your project.

In the following chapter, we will look in more detail at how to control appliances from your Raspberry Pi, such as LEDs, lamps, and other appliances you could have in your home.

4
Controlling Appliances from the Raspberry Pi Zero

In previous chapters, we mainly focused on using sensors with our Raspberry Pi, to log data, display it graphically, and also to build a nice thermostat based on the Raspberry Pi Zero.

In this chapter, we are going to focus solely on controlling devices using the Raspberry Pi Zero. Indeed, in any smart home, you are going to want to control devices in order to automate your home.

To cover most of the devices you could have in your home, we are going to see three examples in this chapter. First we will learn how to control and dim a simple LED, which means you'll learn how to control any LED-based lighting in your home. Then, we'll see how to control a DC motor using the Raspberry Pi Zero, which can, for example, be applied to control window blinds. Finally, we'll see how to control a lamp and basically, any on/off appliances in your home. Let's start!

Hardware and software requirements

As always, we are going to start with a list of required hardware and software components for the project.

Except for the Raspberry Pi Zero, you will need some additional components for each of the sections in this chapter.

For the LED controller section, you will need a simple LED and a 330-Ohm resistor.

To control a DC motor, you will need a L293D motor driver IC and, of course, a motor to control. For this purpose, I used a simple 5V DC motor.

For the lamp controller section, the most important component will be the PowerSwitch Tail Kit. This component allows your Pi to control electrical appliances such as lamps, heaters, and other appliances that use mains electricity to function.

Finally, you will need the usual breadboard and jumper wires.

This is the list of components that you will need for this whole chapter, not including the Raspberry Pi Zero:

- LED (`https://www.sparkfun.com/products/9590`)
- 330-Ohm resistor (`https://www.sparkfun.com/products/11507`)
- L293D motor driver (`https://www.sparkfun.com/products/315`)
- 5V DC motor (`https://www.sparkfun.com/products/11696`)
- Battery pack (`https://www.sparkfun.com/products/9835`)
- PowerSwitch Tail Kit (`https://www.adafruit.com/products/268`)
- Breadboard (`https://www.adafruit.com/products/64`)
- Jumper wires (`https://www.adafruit.com/products/1957`)

On the software side, you will need to install the `pi-gpio` package, which we will later use to dim the LED and to control the speed of the DC motor. To do so, open a terminal on your Pi, and type the following:

```
wget abyz.co.uk/rpi/pigpio/pigpio.zip
```

Then, unzip the archive with the following:

```
unzip pigpio.zip
```

After that, navigate to the unzipped folder with the following:

```
cd PIGPIO
```

Then, build the code using the following:

```
make
```

Finally, install the package on your computer with the following:

```
sudo make install
```

Controlling LEDs

In this first project of the chapter, we are going to see how to control LEDs using your Raspberry Pi Zero. As an example, here we'll see how to control and dim a single LED that we will place on a breadboard. However, the same code can be applied to any kind of LED lighting in your home, or to LED strips.

Let's first see how to assemble this project. Place the LED on the breadboard in series with the 330 Ohm resistor—the longest pin of the LED in contact with the resistor. Then, connect the other side of the resistor to the GPIO18 pin on the Raspberry Pi and the other end of the LED to a GND pin of the Raspberry Pi.

You can, of course, use a cobbler cable kit to easily connect the Pi to the LED. Here, and for the rest of this chapter, I just used two simple jumper wires so you can really see the connections in the images.

This is the final result:

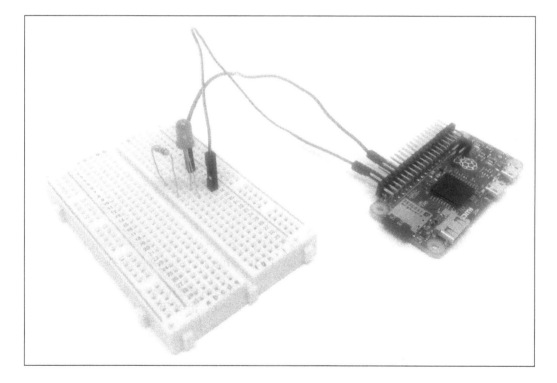

Now that the project is assembled, we are going to test it. To do so, we'll run a simple code that will basically continuously change the intensity of the LED, from 0 to the maximum brightness.

This is the complete code to test our project:

```
// Modules
var Gpio = require('pigpio').Gpio;
// Create led instance
var led = new Gpio(18, {mode: Gpio.OUTPUT});
var dutyCycle = 0;

// Go from 0 to maximum brightness
setInterval(function () {
  led.pwmWrite(dutyCycle);

  dutyCycle += 5;
  if (dutyCycle > 255) {
    dutyCycle = 0;
  }
}, 20);
```

We can already test this code. Make sure to grab the code from this book's GitHub repository, navigate into the folder of this project with a terminal on the Pi, and type the following:

`sudo npm install pigpio`

This will install the required Node.js module to control the LED. Then, type the following:

`sudo node led_test.js`

You should immediately see the LED gradually going from completely off to full brightness, meaning we can indeed dim an LED using our Raspberry Pi Zero board!

This is great, but we can do better. We are now going to see how to dim the LED using a graphical interface, in which you'll be able to control the intensity of the LED using a slider.

As for the thermostat project, we are going to use Node.js again here, along with an HTML interface and some JavaScript to link the interface to the Node.js server.

Let's first have a look at the Node.js code. We'll again use the `Express` module to structure our app, along with the `pigpio` module we used earlier to dim the LED:

```
// Modules
var Gpio = require('pigpio').Gpio;
var express = require('express');

// Express app
var app = express();

// Use public directory
app.use(express.static('public'));
```

Then, we define an object that will allow us to control the state of the LED:

```
var led = new Gpio(18, {mode: Gpio.OUTPUT});
```

We can then define the routes of our app. The first one is to serve the interface when we access it via a web browser:

```
app.get('/', function (req, res) {

  res.sendfile(__dirname + '/public/interface.html');

});
```

Then, we also create a route to set the intensity of the LED:

```
app.get('/set', function (req, res) {

  // Set LED
  dutyCycle = req.query.dutyCycle;
  led.pwmWrite(dutyCycle);

  // Answer
  answer = {
    dutyCycle: dutyCycle
  };
  res.json(answer);

});
```

Finally, we start the Node.js server with the following:

```
app.listen(3000, function () {
  console.log('Raspberry Pi Zero LED control started!');
});
```

Let's now have a look at the HTML interface. It starts by including modules such as jQuery and Bootstrap:

```
<head>
  <script src="https://code.jquery.com/jquery-2.2.4.min.js"></script>
  <link rel="stylesheet" href="https://maxcdn.bootstrapcdn.com/
bootstrap/3.3.6/css/bootstrap.min.css">
  <script src="https://maxcdn.bootstrapcdn.com/bootstrap/3.3.6/js/
bootstrap.min.js"></script>
  <script src="js/interface.js"></script>
  <link rel="stylesheet" href="css/style.css">
  <meta name="viewport" content="width=device-width, initial-scale=1">
</head>
```

Then, in the body of the interface, we simply define a slider element, which we will use to control the LED:

```
<body>

<div id="container">

  <h3>LED Control</h3>

  <div class='row'>

    <div class='col-md-4'></div>
    <div class='col-md-4 text-center'>
     <input id="duty-cycle" type="range" value="0" min="0" max="255"
step="1">
    </div>
    <div class='col-md-4'></div>

  </div>

</div>

</body>
```

Finally, in the file called `script.js`, we link the slider element to the Node.js server so it automatically sets the intensity of the LED whenever we use the slider:

```
$( "#duty-cycle" ).mouseup(function() {

    // Get value
    var dutyCycle = $('#duty-cycle').val();

    // Set new value
    $.get('/set?dutyCycle=' + dutyCycle);

});
```

It's now finally time to test our application! First, grab all the code from this book's GitHub repository and navigate to the folder of the project like before. Then, install Express with the following command:

```
sudo npm install express
```

When this is done, start the server with the following command:

```
sudo node led_control.js
```

You can now test the project by entering the following command in your browser (replacing the IP address with the one for your Pi):

```
http://192.168.0.103:3000/set?dutyCycle=20
```

You should immediately see the LED dimmed and also get the confirmation in your browser:

Then, access the interface directly by typing the following URL:

```
http://192.168.0.103:3000
```

You should immediately see a very basic interface with a slider:

You can now try it: as soon as you release the mouse from the slider, you should see that the LED is instantly dimmed to the value you set with the slider! You can now connect this project to any LED-based lighting in your home (that uses DC current), and start controlling it from a nice interface using your Raspberry Pi Zero!

Controlling the speed of a DC motor

In any smart home, chances are you will find a DC motor somewhere that you will need to automate. It could, for example, be on electric window blinds, or on an automated garage door. In this section, we are going to see how to control the speed of a simple DC motor and you will then be able to apply this to any motors you already have in your home.

Let's first see how to connect the DC motor to your Raspberry Pi Zero board. We actually won't connect the motor directly to the Raspberry Pi, as this would require a lot of external components, such as transistors, diodes, and so on. Instead, we'll use the L293D chip, which is a dedicated IC to control DC motors.

First, place the L293D on the board. The following diagram shows the pinout of the L293D:

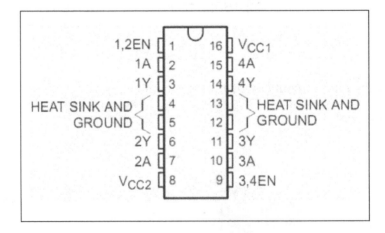

You basically need to connect the components to the L293D as follows:

- GPIO14 of the Raspberry Pi to pin 1A
- GPIO15 of the Raspberry Pi to pin 2A
- GPIO18 of the Raspberry Pi to pin 1,2EN
- DC motor to pin 1Y and 2Y
- 5V of the Raspberry Pi to VCC1
- GND of the Raspberry Pi to GND
- Battery pack to VCC2 and GND

The following image shows the final result:

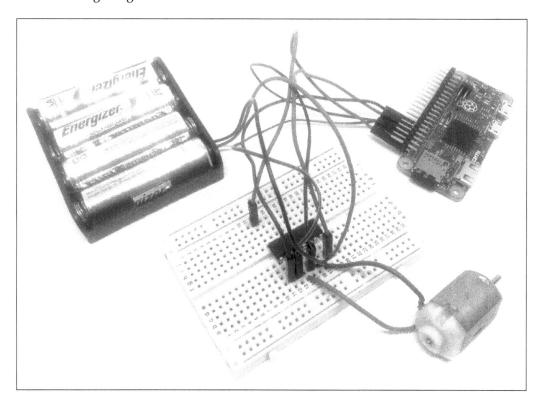

We are now going to see how to perform a simple test of the DC motor to see if it is working correctly. We are simply going to make the motor accelerate from null speed to its maximum speed. The following is the code to do exactly that:

```
// Modules
var Gpio = require('pigpio').Gpio;

// Create motor instance
var motorSpeed = new Gpio(18, {mode: Gpio.OUTPUT});
var motorDirectionOne = new Gpio(14, {mode: Gpio.OUTPUT});
var motorDirectionTwo = new Gpio(15, {mode: Gpio.OUTPUT})

// Init motor direction
motorDirectionOne.digitalWrite(0);
motorDirectionTwo.digitalWrite(1);

var dutyCycle = 0;

// Go from 0 to maximum speed
setInterval(function () {
  motorSpeed.pwmWrite(dutyCycle);

  dutyCycle += 5;
  if (dutyCycle > 255) {
    dutyCycle = 0;
  }
}, 20);
```

You can now save this code in a JavaScript file, and use a Terminal to navigate to the folder where this file is located. Then, type the following:

`sudo npm install pigpio`

This will install the required module to use this test file. Then, launch the test with the following:

`sudo node motor_test.js`

You should immediately see the motor going from zero to its maximum speed, and then start again.

Just like the LED previously, we want to be able to control the speed of the motor using a graphical interface. As the code for this part is really similar to the code for the previous section, I will only highlight the main differences here.

First, we create instances of the GPIO module to control each output pin we need:

```
var motorSpeed = new Gpio(18, {mode: Gpio.OUTPUT});
var motorDirectionOne = new Gpio(14, {mode: Gpio.OUTPUT});
var motorDirectionTwo = new Gpio(15, {mode: Gpio.OUTPUT});
```

We also define a route to set the speed of the motor:

```
app.get('/set', function (req, res) {

  // Set motor speed
  speed = req.query.speed;
  motorSpeed.pwmWrite(speed);

  // Set motor direction
  motorDirectionOne.digitalWrite(0);
  motorDirectionTwo.digitalWrite(1);

  // Answer
  answer = {
    speed: speed
  };
  res.json(answer);

});
```

For now, we'll set the direction inside the code for convenience, but you can of course change it now.

As for the LED control, we'll use an interface that will just display a single slider to control the speed of the motor.

It's now time to test the project! Grab all the code from this book's GitHub repository and inside the folder where you extracted the code, type the following:

```
sudo npm install express
```

This will install the Express module that is required for the project. Then, you can start the server using the following:

```
sudo node motor_control.js
```

You can now navigate to the main interface of the project (of course replacing the IP with the one for your Pi):

```
http://192.168.0.103:3000
```

You should see an interface displaying a simple slider to control the speed of the motor:

You can now try it: moving the slider and releasing the mouse should immediately change the speed of the motor to the desired speed. You can, of course, now also add a switch to the interface to also change the direction of the motor using this simple graphical interface. This will allow us to control a motor for which you might want to reverse the direction of rotation, for example, to open or close a garage door.

Controlling home appliances

In the final section of this chapter, we are going to see how to control appliances in your home that can only be set to on or off, for example, lamps, but also heaters, coffee machines, and other appliances. In this section, you are going to learn how to control a simple desk lamp using your Raspberry Pi Zero.

Let's first see how to assemble the project. Simply connect the Vin+ pin of the PowerSwitch Tail Kit to the GPIO18 pin on the Raspberry Pi Zero, and the two remaining pins of the PowerSwitch Tail to GND.

The following image shows the final result:

Of course, after this you need to connect a lamp to the project. For that, I used a simple 30W desk lamp. You simply need to connect the appliance you want to control to the female plug of the PowerSwitch Tail, and then connect it to the mains electricity via the male power plug.

We are now going to see how to control the lamp using a simple interface, which will run on our Raspberry Pi. For that, we'll again use Node.js, along with an interface written in HTML.

Let's first have a look at the Node.js file. We include the Express module, and also define to which pin the PowerSwitch Tail is connected:

```
// Modules
var express = require('express');

// Express app
var app = express();

// Pin
var lampPin = 12;

// Use public directory
app.use(express.static('public'));
```

Then, we define a main route for the interface:

```
app.get('/', function (req, res) {

  res.sendfile(__dirname + '/public/interface.html');

});
```

We also define a route that we will use to turn the lamp on. Here, we are again going to use the aREST framework, which will allow us to easily control the Raspberry Pi. The following is the complete code for this route:

```
app.get('/on', function (req, res) {

  piREST.digitalWrite(lampPin, 1);

  // Answer
  answer = {
    status: 1
  };
  res.json(answer);

});
```

We also do the same for the off route:

```
app.get('/off', function (req, res) {

  piREST.digitalWrite(lampPin, 0);
```

```
// Answer
answer = {
  status: 0
};
res.json(answer);

});
```

Finally, we initialize the aREST instance and start the server:

```
// aREST
var piREST = require('pi-arest')(app);
piREST.set_id('34f5eQ');
piREST.set_name('my_rpi_zero');

// Start server
app.listen(3000, function () {
  console.log('Raspberry Pi Zero lamp control started!');
});
```

Let's now have a look at the HTML interface. It simply consists of two buttons, one to turn the light on, and one to turn it off again:

```
<div class='row'>

    <div class='col-md-4'></div>
    <div class='col-md-2'>
      <button id='on' class='btn btn-block btn-primary'>On</button>
    </div>
    <div class='col-md-2'>
      <button id='off' class='btn btn-block btn-warning'>Off</button>
    </div>
    <div class='col-md-4'></div>

</div>
```

Inside the `script` file, we also link each of the buttons to the corresponding action on the server:

```
$( "#on" ).click(function() {

    // Set lamp ON
    $.get('/on');

});
```

```
$( "#off" ).click(function() {

    // Set lamp OFF
    $.get('/off');

});
```

It's now finally time to test the project! Make sure that a lamp is connected to the PowerSwitch Tail and that it is turned on. Also, make sure the project is connected to the mains electricity.

Then, grab all the code from this book's GitHub repository and place it in a folder. Then, navigate to the `lamp_control` folder, and type the following inside a Terminal:

```
sudo npm install express pi-arest
```

This will install all the required components for the project. After that, type the following:

```
sudo node lamp_control.js
```

This will start the server on the Pi. After that, navigate to the following URL by replacing the IP address with the one for your Pi:

```
http://192.168.0.103:3000
```

You should see the interface that we just created, with two push buttons:

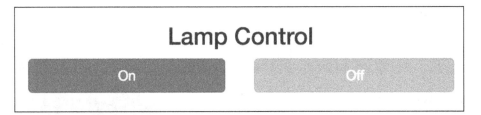

You can now try the interface. When you press on the button, it should immediately set the correct state to the lamp or any appliance that you connected to the PowerSwitch Tail.

 In case it doesn't work, first make sure that you set the correct pin inside the code, which corresponds to the pin where you connected the PowerSwitch Tail on the Raspberry Pi. Also check that the lamp is actually turned on in cases where there is a physical switch on the lamp.

Summary

In this chapter, we saw how to control devices from the Raspberry Pi Zero board, which is critical in any smart home that you want to automate. First we looked at how to control and dim an LED, which you can use to control LED-based lighting and LED strips in your home. We also looked at how to control the speed of a DC motor, which you can apply to control the motor of a garage door, for example. Finally, we saw how to control any appliances in your home, such as lamps, using a graphical interface running on your Pi.

You can, of course, already use what you learned in this chapter and adapt it to your own projects, for example, by applying everything you learned in this project to control several devices from the same interface running on your Pi.

In the following chapter, we are going to apply what we learned in this chapter by building a smart energy meter based on the Raspberry Pi Zero.

Making a Smart Plug with the Raspberry Pi Zero

5

You have probably seen those smart plugs in your local shop: they allow you to not only control appliances remotely, but also to measure the energy consumption of the device connected to the plug. These smart plugs are now available nearly everywhere, from well-known brands such as Belkin.

The following image shows one I bought as an experiment a while ago to see what components were inside:

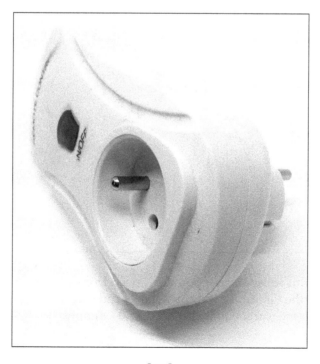

In this chapter, we are going to learn how to use the Raspberry Pi Zero to make a project that has the same functionalities as a smart plug. You will be able to control an electrical device such as a lamp via Wi-Fi, and also measure its electrical consumption in real time.

Because we are building this device ourselves, we will of course be able to customize it for our needs. For example, we'll learn how to log the data measured by the plug in a database so it can be used later. Let's start!

Hardware and software requirements

As always, we are going to start with a list of required hardware and software components for the project.

Except for the Raspberry Pi Zero, you will need some additional components for each of the sections in this chapter.

The most important component will be a current sensor, which we will use to know how much current is flowing through the device. For that, we will use the ECS-1030 non-invasive current sensor. The following is an image of this sensor:

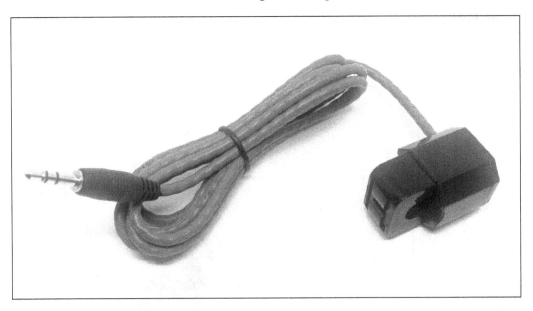

The advantage of this sensor is that you don't need to cut anything to measure the current flowing in the device. To use this sensor, and convert the current it measures to a voltage we can measure, you'll also need a 10-Ohm resistor.

However, we can't directly connect this device to our Raspberry Pi. First, the device has a jack connector at the end, so we need a jack-to-breadboard adapter to connect it first to a breadboard and then to Pi.

Also, we can't connect it to Pi because the Raspberry Pi can only read digital signals. Or, the current sensor is returning an analog signal, which is proportional to the measured current.

To solve this second problem, we'll use a MCP3008 chip, which is an analog-digital converter that can be easily interfaced with the Pi.

To control a device from the Pi, the most important component will be the PowerSwitch Tail Kit. This component allows your Pi to control electrical appliances such as lamps, heaters, and other appliances that use mains electricity to function.

Finally, you will need the usual breadboard and jumper wires.

The following is the list of components you will need for this whole chapter, not including the Raspberry Pi Zero:

- Non-invasive current sensor (`https://www.sparkfun.com/products/11005`)
- 10-Ohm resistor load
- MCP3008 ADC (`https://www.adafruit.com/product/856`)
- Jack to breadboard adapter (`https://www.sparkfun.com/products/11570`)
- PowerSwitch Tail Kit (`https://www.adafruit.com/products/268`)
- Breadboard (`https://www.adafruit.com/products/64`)
- Jumper wires (`https://www.adafruit.com/products/1957`)

To actually have a device to control, I used a standard 15W desk lamp.

Hardware configuration

We are now going to assemble the hardware for this project. The first thing we have to do is to connect the current sensor to the PowerSwitch Tail so we can measure the current flowing into the device connected to the smart plug. For that, you'll need to expose a bare cable coming from the PowerSwitch Tail. Then, open the current sensor and close it firmly around this cable.

The following image shows how the connection should look:

Now, we are going to connect the MCP3008 to the Raspberry Pi. To help you out, the following figure shows the connector of the Raspberry Pi Zero with the numbers of the GPIO pins:

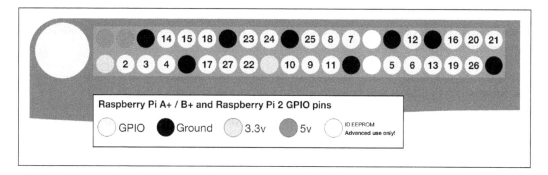

Also, you will need to know the pins of the MCP3008 chip:

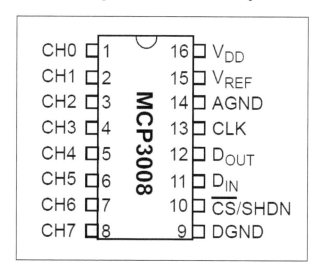

Let's now connect the MCP3008 to the Raspberry Pi, using the following connections:

- VDD and VREF to the 3.3V pin of the Raspberry Pi
- DGND and AGND to the GND pin of the Raspberry Pi
- CLK to GPIO 11 of the Raspberry Pi
- DOUT to GPIO 9 of the Raspberry Pi
- DIN to GPIO 10 of the Raspberry Pi
- CS to GPIO 8 of the Raspberry Pi

After that, plug the current sensor into the Jack adapter and place the adapter on the breadboard. Then, connect the 10-Ohm resistor in series with the SLEEVE and TIP pins on the adapter.

Now, connect one side of this resistor to the GND on the breadboard (for example, to the Raspberry Pi GND) and the other side to channel 5 of the MCP3008 chip.

Finally, connect the PowerSwitch Tail to the Pi: connect the Vin+ pin to GPIO 18 and the two other pins to the GND.

The following image shows the final result:

Also, connect the device you want to control, for example, a lamp, to the female plug of the PowerSwitch. You can now again plug your Pi into the power source and the PowerSwitch Tail to the mains electricity.

Configuring the smart plug

We are now going to configure the Raspberry Pi so it behaves like a smart plug. As usual, we'll use Node.js to code the software that will control our Raspberry Pi Zero board.

We start by importing all the required modules for the project:

```
var mcpadc = require('mcp-spi-adc');
var express = require('express');
var app = express();
var piREST = require('pi-arest')(app);
```

Note that we are using the `mcp-spi-adc` module here, which will allow us to easily read data from the MCP3008 chip.

Next, we define the channel to which the current sensor is connected:

```
var channel = 5;
```

We also set the value of the load resistance we are using for the sensor:

```
var resistance = 10;
```

This will allow us to calculate the actual current flowing through the sensor later on.

So far, you might have noticed that we don't measure the voltage in this project. Indeed, even if we could add another circuit to measure the voltage, we can simply set it in the code:

```
var voltage = 230; // Europe
```

Note that you will need to change that to the voltage used in your country, for example, 110V in the US.

As for the other projects in this book, we are using the aREST framework to control our Pi remotely. We need to initialize the module by giving a name and ID to our project:

```
piREST.set_id('34f5eQ');
piREST.set_name('energy_meter');
piREST.set_mode('bcm');
```

After that, we start the server on port 80:

```
var server = app.listen(80, function() {
    console.log('Listening on port %d', server.address().port);
});
```

We still need to actually measure data from the sensor. This is the code that will measure data from the sensor every 500 ms:

```
var sensor = mcpadc.open(channel, {speedHz: 20000}, function (err) {
    if (err) throw err;

    // Measurement interval
    setInterval(function () {

        // Read
        sensor.read(function (err, reading) {
            if (err) throw err;

            // Calculate current
            var measuredVoltage = reading.value * 3.3;
            var measuredCurrent = (measuredVoltage/resistance) * 2000 /
                1.41;

            // Calculate power
            var power = voltage * measuredCurrent;

            // Assign to aREST
            piREST.variable('power', power.toFixed(2));
            piREST.variable('current', measuredCurrent.toFixed(2));

            // Log output
            console.log("Measured current: " + measuredCurrent.toFixed(2) +
    'A');
            console.log("Measured power: " + power.toFixed(2) + 'W');

        });
    }, 500);
});
```

Let's see what this code does. We first measure the voltage at the analog-digital converter, which is proportional to the current flowing through the device of our smart plug. We need to multiply this reading by 3.3 to get a voltage from the value returned by the analog-digital converter.

Then, to get the current, we first need to divide the voltage by the value of the resistance. Then, we need to multiply it by 2000, which is the ratio between the current induced in the sensor and the current actually flowing through the device. Finally, we need to divide the result by 1.41, to get the effective value of the current.

To get the power, we just multiply the measured current by the voltage.

Finally, we also expose those two measured values to the aREST API so we can access them later.

It's finally time to test the project! Get all the files from this book's GitHub and put them inside a folder in your Pi. Make sure to modify the file called `meter.js` to change the value of the resistance, in case you are using a different value.

Then, install the required modules from a terminal with the following command:

```
sudo npm install express pi-arest mcp-spi-adcsdfsd
```

Once that's done, start the project with the following command:

```
sudo node meter.js
```

You should immediately see the measurements in the console, showing a null current and power, as the device is currently off:

```
Measured current: 0.00A
Measured power: 0.00W
Measured current: 0.00A
Measured power: 0.00W
```

Let's now switch the device on to see if the current and power measurements are working correctly. For that, first get the IP address of your Pi using the `ifconfig` command. Let's assume for the rest of this chapter that it is 192.168.0.105.

Go to your favorite web browser and type the following:

```
http://192.168.0.105/digital/18/1
```

This should immediately switch the device connected to the project on, for example, the desk lamp I connected to the PowerSwitch Tail. You should see the current and power measurements in the console:

```
Measured current: 0.07A
Measured power: 15.03W
Measured current: 0.07A
Measured power: 15.03W
```

As you can see, the sensor is quite precise, as I obtained a 15.03W reading for my 15W desk lamp. Note that the current sensor can measure up to 30A, so make sure you are using a device that is working with a current smaller than this limit.

Creating an interface for the smart plug

Commercial smart plugs usually come with a nice interface, which you can use from your phone or computer to control the plug via Wi-Fi. In this section, we are going to do exactly the same: build a simple interface that we will use to control the device connected to the smart plug, and also visualize the current and power consumption of the device.

As the code for this part is quite similar to the code of the previous section, I will only highlight the differences here.

Inside the Node.js JavaScript file, we declare the public folder in which we will store the interface:

```
app.use(express.static('public'));
```

Then, we need to declare to which pin we connected the output of the smart plug:

```
var outputPin = 18;
```

Using Express, we can now define some routes. We define the main route of the application to redirect to the interface file:

```
app.get('/', function (req, res) {

  res.sendfile(__dirname + '/public/interface.html');

});
```

Then, as we saw in the previous chapter, we declare two routes to control the output of the project: one to switch the device on, and one to switch it off:

```
app.get('/on', function (req, res) {

  piREST.digitalWrite(outputPin, 1);

  // Answer
  answer = {
    status: 1
  };
  res.json(answer);

});

app.get('/off', function (req, res) {

  piREST.digitalWrite(outputPin, 0);

  // Answer
  answer = {
    status: 0
  };
  res.json(answer);

});
```

Let's now see the files for the interface. There will be one HTML file, which contains the elements of the interface, and one JavaScript file to make the link between the elements and the Node.js software.

Let's start with the HTML file. We need to define the two buttons that we will use to control the device:

```
<div class='row'>

    <div class='col-md-4'></div>
    <div class='col-md-2'>
      <button id='on' class='btn btn-block btn-primary'>On</button>
    </div>
    <div class='col-md-2'>
      <button id='off' class='btn btn-block btn-warning'>Off</button>
    </div>
    <div class='col-md-4'></div>

</div>
```

Then, we also define two indicators, called **current** and **power**, which will contain the values measured by the project:

```
<div class='row'>

  <div class='col-md-4'></div>
  <div class='col-md-4'>
    Current consumption: <span id='current'></span> A
  </div>
  <div class='col-md-4'></div>

</div>

<div class='row'>

  <div class='col-md-4'></div>
  <div class='col-md-4'>
    Power consumption: <span id='power'></span> W
  </div>
  <div class='col-md-4'></div>

</div>
```

Let's now see the content of the JavaScript file. First we make the link between the buttons and the Node.js server by calling the correct action when a button is pressed:

```
$( "#on" ).click(function() {

  // Set lamp ON
  $.get('/on');

});

$( "#off" ).click(function() {

  // Set lamp OFF
  $.get('/off');

});
```

Then, we define this loop to automatically grab the measurements from the Pi and update the indicators in the interface every second:

```
setInterval(function () {

    // Current
    $.get('/current', function(data) {
      $( "#current" ).text(data.current);
    });

    // Power
    $.get('/power', function(data) {
      $( "#power" ).text(data.power);
    });

}, 1000);
```

It's now time to test the interface! If you followed the instructions from the previous section, you just need to go once more to the folder where you put the project files and type the following:

`sudo node meter_interface.js`

Now, using your favorite browser, go to the IP address of the Pi, for example:

```
http://192.168.0.105/
```

You should see the interface showing the current measurements taken by the board, which should be at zero, as you just started the software:

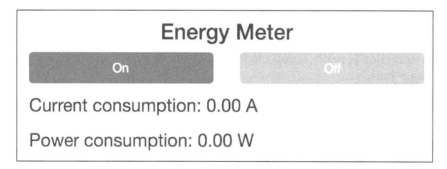

You can now click on the **On** button. You should immediately see the device connected to the project turning on and you should also see the current readings made by the smart plug:

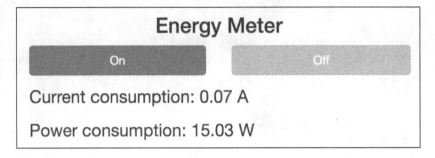

Congratulations, you now have a nice interface that you can use to control your smart plug remotely! Of course, you could also use this interface from a phone or tablet that is connected to the same Wi-Fi network as your Pi.

Logging your energy consumption over time

For now, we built a smart plug that has more or less the same features as a commercial smart plug: it can control a device, measure the power consumption of this device, and also comes with a nice graphical interface. In this section, we are going to go further, and see how we can easily add functions to our project with some lines of code.

As an example, we are going to see how to log the measurements made by the board into a database on the Pi so that those measurements can be recalled later. As the code for this section is really similar to the previous section, I will only highlight the main changes here.

Start by importing the required module for the database:

```
var Datastore = require('nedb')
  db = new Datastore();
```

After that, we define a route to get all the data currently present inside the database:

```
app.get('/data', function (req, res) {

  db.find({}, function (err, docs) {
```

```
    res.json(docs);

  });

});
```

Inside the measurement loop, we create a new set of data at every iteration and store it in the database:

```
var data = {
        current: measuredCurrent.toFixed(2),
        power: power.toFixed(2),
        date: new Date()
    };
    db.insert(data, function (err, newDoc) {
        console.log(newDoc);
    });
```

Let's now try this new piece of software. Again, navigate to the folder where you put the files for this chapter and type the following command:

```
sudo npm install nedb
```

This will install the required module for the database. Then, launch the software with the following command:

```
sudo node meter_log.js
```

You should see the results from the measurements inside the console just as before, but this time with the confirmation that the document was stored in the database:

```
Measured current: 0.07A
Measured power: 15.03W
{ current: '0.07',
  power: '15.03',
  date: Fri Aug 05 2016 09:00:36 GMT+0000 (UTC),
  _id: 'EzLofvbzXK5tRGoW' }
```

Note that I used quite a high refresh rate inside the code for demonstration purposes. Of course, I invite you to modify that in order to avoid filling your database with measurements.

You can also try to read the data that was logged inside the database by going to the following URL:

```
http://192.168.0.105/data
```

You should immediately see the results inside the browser:

← → C ⓘ 192.168.0.105/data   ☆

[{"current":"0.07","power":"15.03","date":"2016-08-05T09:01:24.843Z","_id":"31xTER1HUuD7bXIt"},{"current":"0.00","power":"0.00","date":"2016-08-
05T09:01:19.312Z","_id":"3ZN5RDHW6PdzDppS"},{"current":"0.07","power":"15.03","date":"2016-08-05T09:01:15.802Z","_id":"4Xh0lO7qQIP4H3bL"},
{"current":"0.00","power":"0.00","date":"2016-08-05T09:01:27.860Z","_id":"5wlTripVs6ID8trK"},{"current":"0.07","power":"15.03","date":"2016-08-
05T09:01:20.820Z","_id":"6Cv8nivXWvcaCAVj"},{"current":"0.07","power":"15.03","date":"2016-08-05T09:01:16.300Z","_id":"7tIP1VuVdk340qvn"},
{"current":"0.00","power":"0.00","date":"2016-08-05T09:01:21.826Z","_id":"84yjm9yATLtvkbnx"},{"current":"0.00","power":"0.00","date":"2016-08-
05T09:01:29.368Z","_id":"9y0zgD7NJMd4MRNN"},{"current":"0.00","power":"0.00","date":"2016-08-05T09:01:23.838Z","_id":"IxoddyQeOX5QuVQy"},
{"current":"0.07","power":"15.03","date":"2016-08-05T09:01:25.346Z","_id":"JaeEigcJsCoGMtQA"},{"current":"0.07","power":"15.03","date":"2016-08-
05T09:01:28.363Z","_id":"JjKugnE8og5OagcT"},{"current":"0.07","power":"15.03","date":"2016-08-05T09:01:21.322Z","_id":"KWJuXN0arLrvFDu6"},
{"current":"0.00","power":"0.00","date":"2016-08-05T09:01:15.296Z","_id":"LXPuzhx6eGEnMInW"},{"current":"0.00","power":"0.00","date":"2016-08-
05T09:01:19.815Z","_id":"MuiqP8pnPvlyccki"},{"current":"0.00","power":"0.00","date":"2016-08-05T09:01:22.831Z","_id":"Nn38h7jPE1HAd5X6"},
{"current":"0.00","power":"0.00","date":"2016-08-05T09:01:18.309Z","_id":"PJJIquVvLrrqloe0K"},{"current":"0.00","power":"0.00","date":"2016-08-
05T09:01:17.807Z","_id":"Qfb05b2qRnTNGpqm"},{"current":"0.00","power":"0.00","date":"2016-08-05T09:01:29.869Z","_id":"TAdqpTPGLhNZuPo7"},
{"current":"0.00","power":"0.00","date":"2016-08-05T09:01:22.329Z","_id":"UoOwwFRIWCak0WMd"},{"current":"0.00","power":"0.00","date":"2016-08-
05T09:01:14.798Z","_id":"Z3WyKYEcpXTYBIg1"},{"current":"0.07","power":"15.03","date":"2016-08-05T09:01:20.318Z","_id":"aDkVOaCXnVKEObm0"},
{"current":"0.00","power":"0.00","date":"2016-08-05T09:01:26.352Z","_id":"dWa52MypXOqcWBJ2"},{"current":"0.00","power":"0.00","date":"2016-08-
05T09:01:17.305Z","_id":"el0aoJOtukI3Lq2S"},{"current":"0.00","power":"0.00","date":"2016-08-05T09:01:25.849Z","_id":"g7XOm7PqRhgFHkqT"},
{"current":"0.00","power":"0.00","date":"2016-08-05T09:01:27.357Z","_id":"m6f5xRK0IAIafxjU"},{"current":"0.00","power":"0.00","date":"2016-08-
05T09:01:16.803Z","_id":"t0fmLoD6FEKC3xih"},{"current":"0.00","power":"0.00","date":"2016-08-05T09:01:26.854Z","_id":"tiIfpFANsEzZW9xm"},
{"current":"0.00","power":"0.00","date":"2016-08-05T09:01:18.811Z","_id":"uDLbHD89YW6f9mPi"},{"current":"0.07","power":"15.03","date":"2016-08-
05T09:01:24.341Z","_id":"w7JnaTiLD89kECKL"},{"current":"0.00","power":"0.00","date":"2016-08-05T09:01:23.335Z","_id":"wHHaciHY3TMwehDe"},
{"current":"0.07","power":"15.03","date":"2016-08-05T09:01:28.866Z","_id":"z4bGaH8S6lOa0JiT"}]

You can now use this data for your own applications. For example, it can easily be used to calculate the average daily or monthly energy consumption of the device.

Summary

In this chapter, we learned how to reproduce a smart plug using the Raspberry Pi Zero. We built a device that can control electrical devices and also measure their energy consumption. We built a nice interface to control this device, and also made it log data on the Pi itself.

You can, of course, now improve this project in many ways. You could, for example, plot the data measured by the project and have a real-time graph of the energy consumption of the device. For the more adventurous, you could also think about integrating all the components into a nice 3D-printed case, making it almost like a commercial smart plug.

In the following chapter, we are going to dive into an amazing field: the Internet of Things. We'll see how to use your Pi to send you all kinds of notifications about what is going on in your home.

6

Sending Notifications using Raspberry Pi Zero

In this chapter, we are going to start diving into a very interesting field that will change the way we interact with our environment: the **Internet of Things** (**IoT**). The IoT basically proposes to connect every device around us to the Internet, so we can interact with them from anywhere in the world.

Within this context, a very important application is to receive notifications from your devices when they detect something in your home, for example a motion in your home or the current temperature. This is exactly what we are going to do in this chapter, we are going to learn how to make your Raspberry Pi Zero board send you notifications via text message, email, and push notifications. Let's start!

Hardware and software requirements

As always, we are going to start with the list of required hardware and software components for the project.

Except Raspberry Pi Zero, you will need some additional components for each of the sections in this chapter.

For the first project of this chapter, we are going to use a simple PIR motion sensor to detect motion from your Pi.

Then, for the last two projects of the chapter, we'll use the DHT11 sensor that we have already used in previous chapters.

Finally, you will need the usual breadboard and jumper wires.

This is the list of components that you will need for this whole chapter, not including the Raspberry Pi Zero:

- PIR motion sensor (`https://www.sparkfun.com/products/13285`)
- DHT11 sensor + 4.7k Ohm resistor (`https://www.adafruit.com/products/386`)
- Breadboard (`https://www.adafruit.com/products/64`)
- Jumper wires (`https://www.adafruit.com/products/1957`)

On the software side, you will need to create an account on IFTTT, which we will use in all the projects of this chapter. For that, simply go to:

`https://ifttt.com/`

You should be redirected to the main page of IFTTT where you'll be able to create an account:

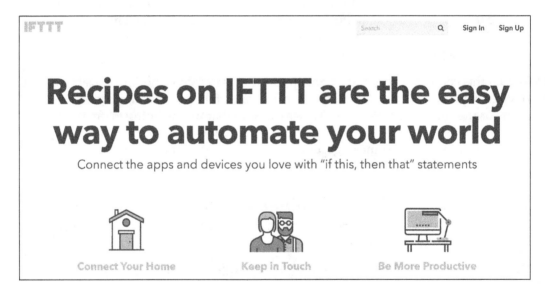

Making a motion sensor that sends text messages

For the first project of this chapter, we are going to attach a motion sensor to the Raspberry Pi board and make the Raspberry Pi Zero send us a text message whenever motion is detected. For that, we are going to use IFTTT to make the link between our Raspberry Pi and our phone. Indeed, whenever IFTTT will receive a trigger from the Raspberry Pi, it will automatically send us a text message.

Lets first connect the PIR motion sensor to the Raspberry Pi. For that, simply connect the VCC pin of the sensor to a 3.3V pin of the Raspberry Pi, GND to GND, and the OUT pin of the sensor to GPIO18 of the Raspberry Pi.

This is the final result:

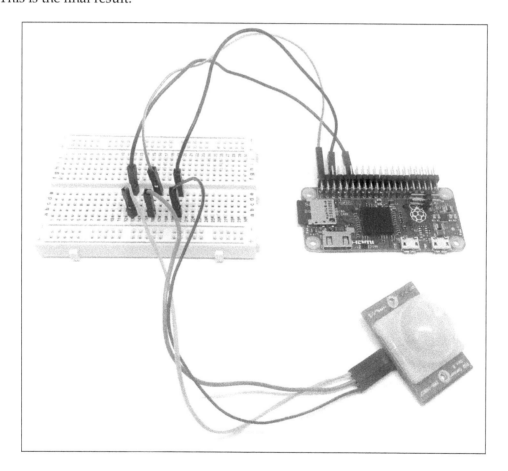

Let's now add our first channel to IFTTT, which will allow us later to interact with the Raspberry Pi and with web services. You can easily add new channels by clicking on the corresponding tab on the IFTTT website. First, add the **Maker** channel to your account:

This will basically give you a key that you will need when writing the code for this project:

The Maker Channel allows you to connect IFTTT to your personal DIY projects. With Maker, you can connect a Recipe to any device or service that can make or receive a web request (aka webhooks). See how others are using the Maker Channel, or share your own experience at hackster.io.

Connected as: marcoschwartz

Reconnect Channel

Create a New Recipe

Disconnect

How to Trigger Events

Your key is:

dPMHywdahaSxQZICaoqnzHxcQ8vNYsTIk-42gSLAFQP

After that, add the **SMS** channel to your IFTTT account. Now, you can actually create your first recipe. Select the Maker channel as the trigger channel:

Choose Trigger Channel

Showing Channels that provide at least one Trigger. View all Channels

maker

Maker

WeMo Coffeemaker

WeMo Maker

Then, select **Receive a web request**:

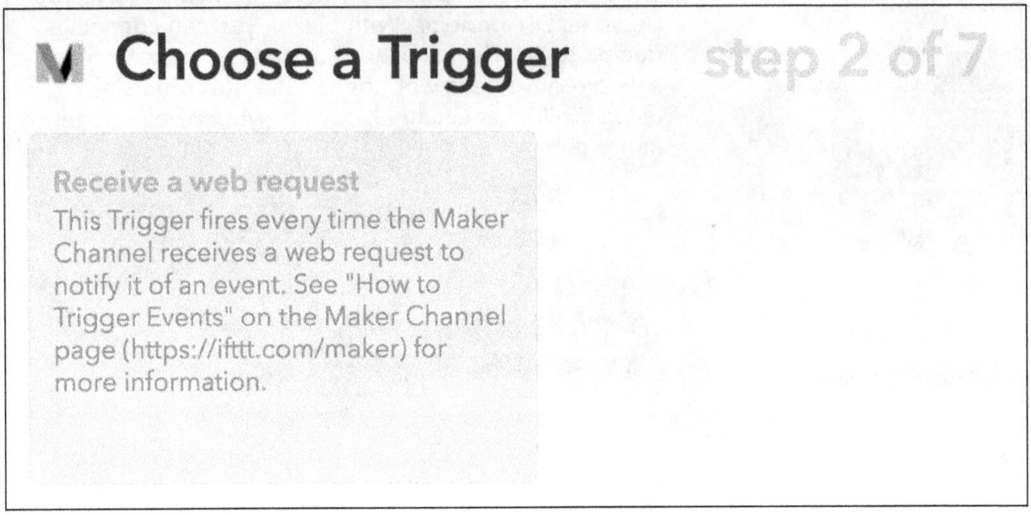

Ⲙ Choose a Trigger   step 2 of 7

Receive a web request
This Trigger fires every time the Maker
Channel receives a web request to
notify it of an event. See "How to
Trigger Events" on the Maker Channel
page (https://ifttt.com/maker) for
more information.

As the name of this request, enter `motion_detected`:

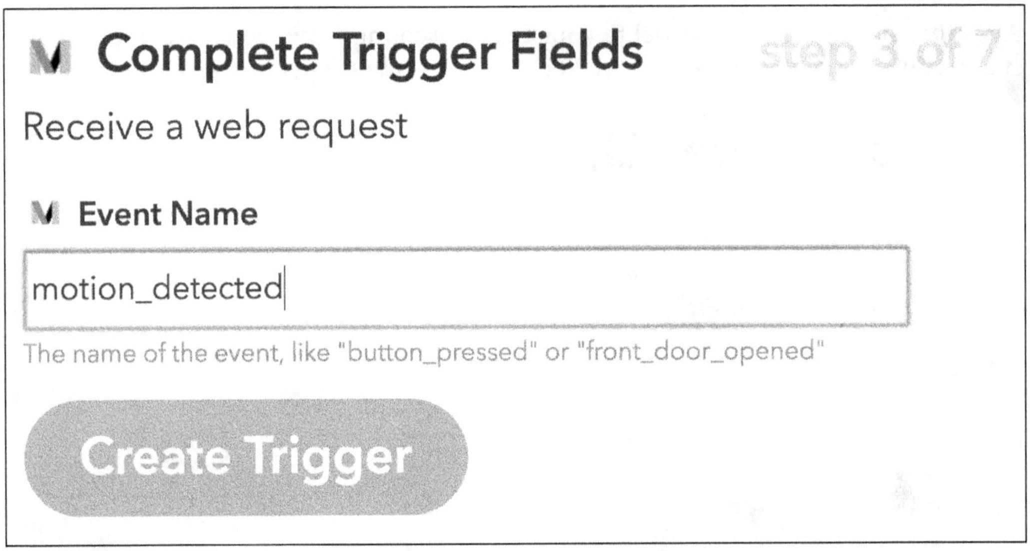

Ⲙ Complete Trigger Fields   step 3 of 7

Receive a web request

Ⲙ Event Name

```
motion_detected
```

The name of the event, like "button_pressed" or "front_door_opened"

Create Trigger

As the action channel, which is the channel that will be executed when a trigger is received, choose the SMS channel:

For the action, choose **Send me an SMS**:

You can now enter the message you want to see in the text messages:

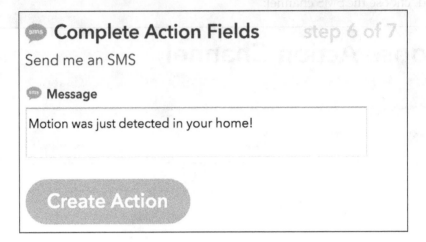

Finally, confirm the creation of the recipe:

Now that our recipe is created and active, we can move on to actually configuring Raspberry Pi so it sends alerts whenever a motion is detected. As usual, we'll use Node.js to code this program.

It starts by including the required modules:

```
// Required modules
var request = require('request');
var gpio = require('rpi-gpio');
```

Then, we define our IFTTT data, which is composed of the Maker key and of the name of the event we want to trigger:

```
// IFTTT data
var key = "your-key";
var eventName = 'motion_detected';
```

Then, we define the pin on which our sensor is connected to:

```
// Motion sensor GPIO
var motionSensorPin = 18;
```

We also need to define a counter that will basically make sure that we don't constantly send alerts to our phone, for example if the sensor stays on for several seconds. For that, we define an interval period of one minute minimum between two alerts:

```
// Counter between two alerts
var interval = 60 * 1000; // 1 minute
var counter = new Date();
```

After that, we configure the `rpi-gpio` module (that we'll use to read data from the sensor) to the BCM configuration scheme, meaning we are using the number of the GPIO of the Raspberry Pi rather than the physical pins:

```
// Setup gpio library
gpio.setMode(gpio.MODE_BCM);
```

Now, every second, we check the status of the sensor (if we didn't trigger an alert already in the last minute):

```
// Check status every second
setInterval(function() {

  // Check counter so we don't trigger alarms all the time
  var currentTime = (new Date()).getTime();
  var counterTime = counter.getTime();
```

```
    if ( (currentTime - counterTime) > interval) {

        // Check sensor
        gpio.setup(motionSensorPin, gpio.DIR_IN, checkSensor);

    }

}, 1000);
```

If some motion is detected, we reset the counter and also send the alert to IFTTT:

```
// Check motion sensor
function checkSensor() {
    gpio.read(motionSensorPin, function(err, value) {

        // If motion is detected, send event to IFTTT
        if (value == true) {

            // Restart Counter
            counter = new Date();

            // Send event
            alertIFTTT();
        }
    });
}
```

This is the function that takes care of sending the data to IFTTT:

```
// Make request
function alertIFTTT() {

    // Send alert to IFTTT
    console.log("Sending alert to IFTTT");
    var url = 'https://maker.ifttt.com/trigger/' + eventName + '/with/
key/' + key;
    request(url, function (error, response, body) {
        if (!error && response.statusCode == 200) {
            console.log("Alert sent to IFTTT");
        }
    });
}
```

It basically uses the request module to send the correct command to IFTTT, passing the name of the event and the key.

We can finally test this first project! Grab the code from the GitHub repository of the book and make sure to modify the code with your own IFTTT key. Then, navigate to the folder where the file is with a terminal and type the following command:

```
sudo npm install rpi-gpio request
```

Once that's done, start the code with:

```
sudo node sms_alerts.js
```

Now, you need to wait at least for the interval time (one minute by default) before the code is active. This will make sure that no motion will be detected when you just start the project for example.

After a minute, pass your hand in front of the sensor: Your Raspberry Pi should immediately send a command to IFTTT and after some seconds you should be able to receive a message on your mobile phone:

To: SMS	Details
	SMS with SMS Today 10:34
Motion was just detected in your home!	

Congratulations, you can now use your Raspberry Pi Zero to send important notifications, on your mobile phone!

 Note that for an actual use of this project in your home, you might want to limit the number of messages you are sending as IFTTT has a limit on the number of messages you can send (check the IFTTT website for the current limit). For example, you could use this for only very important alerts, like in case of an intruder coming in your home in your absence.

Sending temperature alerts through email

In the second project of the chapter, we are going to learn how to send automated email alerts based on data measured by the Raspberry Pi.

Let's first assemble the project. Place the DHT11 sensor on the breadboard and then place the 4.7k Ohm resistor between pin 1 and 2 of the sensor. Then, connect pin 1 of the sensor to the 3.3V pin of the Raspberry Pi, pin 2 to GPIO18, and pin 4 to GND. This is the final result:

Let us now see how to configure the project. Go over to IFTTT and create add the **Email Channel** to your account:

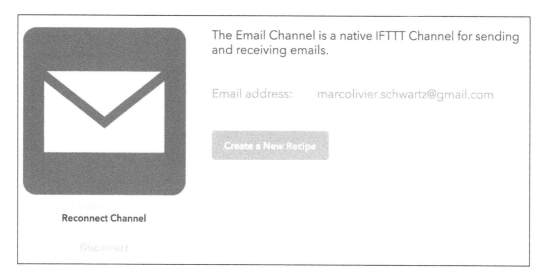

After that, create a new recipe by choosing the Maker channel as the trigger:

For the event, enter `temperature_alert` and then choose **Email** as the action channel:

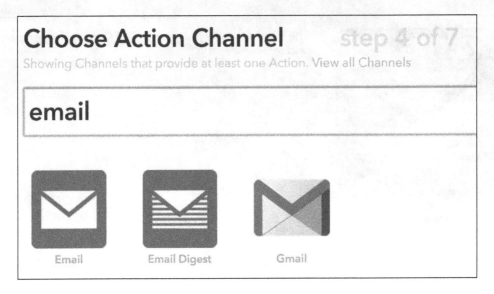

You will then be able to customize the text and subject of the email sent to Pi. As we want to send the emails whenever the temperature in your home gets too low, you can use a similar message:

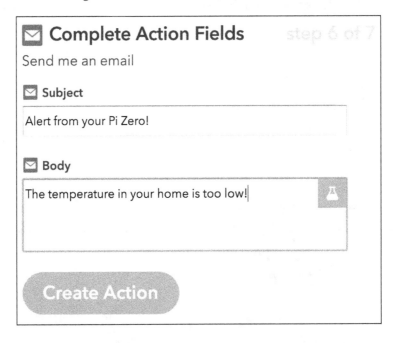

You can now finalize the creation of the recipe and close IFTTT. Let's now see how to configure the Raspberry Pi Zero. As the code for this project is quite similar to the one we saw in the previous section, I will only highlight the main differences here.

It starts by including the required components:

```
var request = require('request');
var sensorLib = require('node-dht-sensor');
```

Then, give the correct name to the event we'll use in the project:

```
var eventName = 'temperature_low';
```

We also define the pin on which the sensor is connected:

```
var sensorPin = 18;
```

As we want to send alerts based on the measured temperature, we need to define a threshold. As it was quite warm when I made this project, I have assigned a high threshold at 30 degrees Celsius, but you can, of course, modify it:

```
var threshold = 30;
```

Then, we initialize the sensor and check the current temperature every 2 seconds:

```
var sensor = {
    initialize: function () {
        return sensorLib.initialize(11, sensorPin);
    },
    read: function () {

        // Read
        var readout = sensorLib.read();
        temperature = readout.temperature.toFixed(2);
        console.log('Current temperature: ' + temperature);

        // Check counter so we don't trigger IFTTT all the time
        var currentTime = (new Date()).getTime();
        var counterTime = counter.getTime();

        if ( (currentTime - counterTime) > interval) {

          if (temperature < threshold) {

            // Restart Counter
            counter = new Date();
```

```
        // Send event
        alertIFTTT();

    }

}

// Repeat
setTimeout(function () {
    sensor.read();
}, 2000);
    }
};
```

It's now time to test the project! Grab the GitHub repository of the book and make sure to modify the code to put your own IFTTT key.

Then, navigate to the folder with the file with a terminal and type the following:

```
sudo npm install request
```

Then, install the module for the DHT sensor with the following command:

```
npm install node-dht-sensor
```

Finally, start the project using the following command:

```
sudo node temperature_alerts.js
```

If like me, you set the threshold to a quite high value, you should quickly receive a message inside your email box:

You can now use your Pi to send automated notifications through email!

Receiving measurement SATA through push notifications

In the last project of this chapter, we'll learn how to use the project we built in the previous section to actually not send you alerts, but just keep you updated about the current temperature and humidity measured by Pi.

Here however, we are going to use something new to alert you: push notifications. These notifications will immediately show up on your phone if you have the right app installed.

As the app, we'll use Pushover that is available for iOS and Android. You can install it from your App Store and find more information at the following URL:

```
https://ifttt.com/
```

Then, add the **Pushover** channel inside IFTTT:

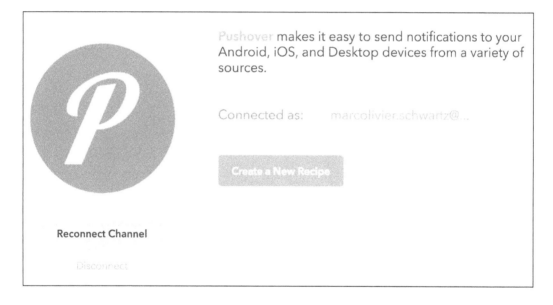

Now, create a new recipe and choose the **Maker** channel again as the trigger:

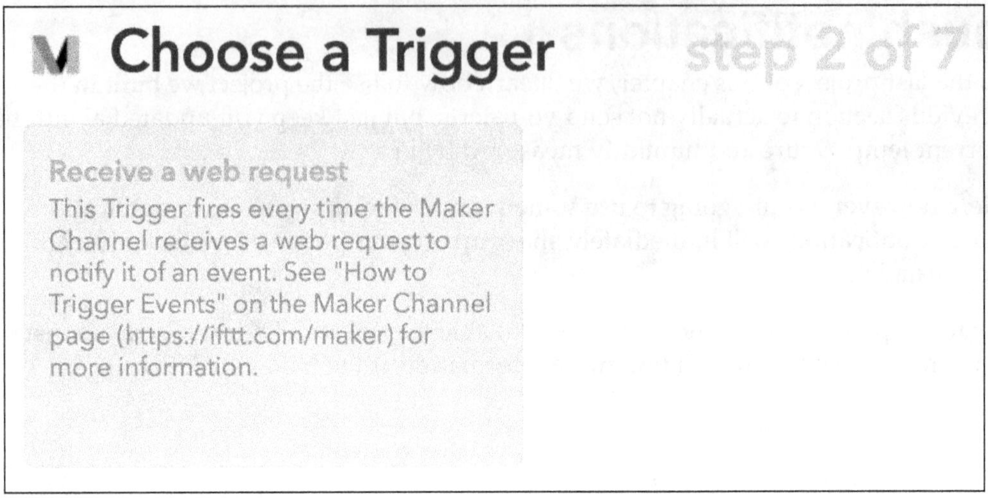

M Choose a Trigger step 2 of 7

Receive a web request
This Trigger fires every time the Maker
Channel receives a web request to
notify it of an event. See "How to
Trigger Events" on the Maker Channel
page (https://ifttt.com/maker) for
more information.

I used **data** as the trigger:

M Complete Trigger Fields step 3 of 7

Receive a web request

M Event Name

data

The name of the event, like "button_pressed" or "front_door_opened"

Create Trigger

Then, select **Pushover** as the action channel and enter the following message:

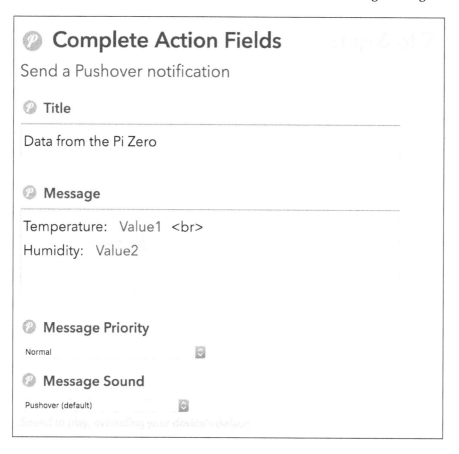

Here, we'll use the variables **Value1** and **Value2** to display the temperature and humidity inside the message. We'll see in a moment how to actually send that to IFTTT from the Pi.

You can also select the target device for this recipe:

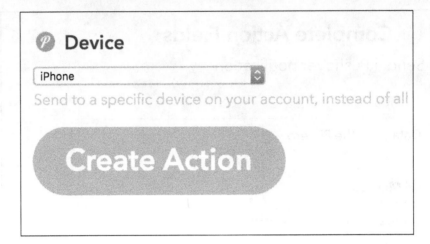

Lets now see how to configure this project. As it's really similar to the previous project, I will just highlight the main differences here.

We need to give the name to the event that we'll trigger from the Pi:

```
var eventName = 'data';
```

Then, inside the sensor object we measure data at regular intervals and pass the measurements to the logIFTTT() function:

```
var sensor = {
    initialize: function () {
        return sensorLib.initialize(11, sensorPin);
    },
    read: function () {

        // Read
        var readout = sensorLib.read();
        temperature = readout.temperature.toFixed(2);
        humidity = readout.humidity.toFixed(2);

        console.log('Current temperature: ' + temperature);
        console.log('Current humidity: ' + humidity);
```

```
        // Send event
        logIFTTT(temperature, humidity);

        // Repeat
        setTimeout(function () {
            sensor.read();
        }, interval);
    }
};
```

Let's now see the details of this function. Compared to the previous projects of this chapter, we are now going to pass the temperature and humidity parameters to the function and send this data to IFTTT:

```
function logIFTTT(temperature, humidity) {

    // Send alert to IFTTT
    console.log("Sending message to IFTTT");
    var url = 'https://maker.ifttt.com/trigger/' + eventName + '/with/
key/' + key;
    url += '?value1=' + temperature + '&value2=' + humidity;
    request(url, function (error, response, body) {
        if (!error && response.statusCode == 200) {
            console.log("Data sent to IFTTT");
        }
    });
}
```

Let's now test the project! First, grab all the code from the GitHub repository of the book and make sure to modify the code with your IFTTT maker key. Then, install the request module with the following command:

```
sudo npm install request
```

Once that's done, install the module for the DHT sensor:

```
npm install node-dht-sensor
```

Finally, start the project using this command:

```
sudo node temperature_notifications.js
```

After a minute, you should get the notification on your phone displaying the current temperature and humidity in your home:

You can now use your Pi to receive automated reports containing data about your home!

Summary

In this chapter, we learned all the basics about sending automated notifications from your Raspberry Pi. We learned, for example, how to send notifications via email, text messages, and push notifications. This is really important to build a smart home, as you want to be able to get alerts in real-time from what's going on inside your home and also receive regular reports about the current status of your home.

You can of course improve the projects of this chapter in many ways. It would be easy for example to have several Pi Zero boards in your home, each sending you alerts on your phone for example. You could give a name to each of the boards and include that name inside the alerts so you know which Pi sent the message.

In the next chapter, we are going to use everything we learned so far in the book to build a simple security system using the Raspberry Pi Zero board.

7
Use the Raspberry Pi Zero to Build a Security System

In this chapter, we are going to learn how to build a modular security system using the Raspberry Pi Zero board. The Raspberry Pi board is really cheap and has a very small form factor, you can use many such boards inside your home to build a complete security system for your home.

We are going to integrate three types of components into our system: motion sensors, alarms, and security camera. These modules will communicate with a central server application that will either run on your computer or on another Raspberry Pi. First, we are going to see how to configure each board individually and then configure the central server and a basic interface.

Hardware and software requirements

As always, we are going to start with the list of required hardware and software components for the project.

In this chapter, we are going to use at least three Raspberry Pi Zero boards: for a motion sensor, an alarm module, and a camera module. Of course, you can perfectly use more of each module in your security system.

For the motion sensor module, I will use a simple PIR motion sensor.

Then, for the alarm module, I will be using a small buzzer, as well as an LED and a 330 Ohm resistor.

For the camera module, I will use a Logitech C270 webcam. Here, any camera compatible with the UVC protocol would work, which is the case for most of the cameras sold these days.

Finally, you will need the usual breadboard and jumper wires.

This is the list of components that you will need for this chapter, not including the Raspberry Pi Zero:

- PIR motion sensor (https://www.sparkfun.com/products/13285)
- LED (https://www.sparkfun.com/products/9590)
- 330 Ohm resistor (https://www.sparkfun.com/products/11507)
- Logitech C270 USB camera (http://www.logitech.com/en-us/product/hd-webcam-c270)
- Breadboard (https://www.adafruit.com/products/64)
- Jumper wires (https://www.adafruit.com/products/1957)

Of course, all the additional components, for example the WiFi dongle and power supply, will need to be multiplied by the number of Raspberry Pi boards that you will use inside the project.

On the software side, you will just need to have Node.js installed on your Raspberry Pi Zero boards.

Building a motion sensor with the Pi Zero

The first module that we are going to assemble in this chapter is the motion sensor module. These modules will be deployed in key parts of your home, to detect any intruder in your home.

The hardware configuration for this part will actually be very simple. First, connect the VCC pin of the motion sensor to a 3.3V pin of the Raspberry Pi. Then, connect the GND pin of the sensor to one GND pin of the Pi. Finally, connect the OUT pin of the motion sensor to the GPIO17 pin of the Raspberry Pi. You can refer to the previous chapters to find out about pin mapping of the Raspberry Pi Zero board.

This is the final result:

Let's now see how to configure this module so we can access it remotely through WiFi. This application will be based on the **aREST framework** again, which we already saw in the previous chapters of the book.

Here is the complete code for this part:

```
// Modules
var express = require('express');

// Express app
var app = express();

// aREST
var piREST = require('pi-arest')(app);
piREST.set_id('34f5eQ');
piREST.set_name('motion_sensor');
piREST.set_mode('bcm');

// Start server
app.listen(3000, function () {
   console.log('Raspberry Pi Zero motion sensor started!');
});
```

You can now simply grab this code from the GitHub repository of the book or simply paste it into a file called `motion_sensor.js`, then using a terminal inside the same folder as the file type:

sudo npm install express pi-arest

Once the required modules are installed, type the following command to start the project:

sudo node motion_sensor.js

Finally, navigate to the IP address of your Pi on port `3000`, followed by the digital command on pin 17:

http://192.168.0.105:3000/digital/17

This should immediately return a JSON object with the value of pin 17. You can now try to pass your hand in front of the sensor and repeat the operation: you should be able to immediately see that the value of pin 17 is equal to 1, indicating that motion has been detected by the sensor.

Making a simple alarm module

In the second part of this chapter, we are going to learn how to build an alarm module for our security system. You will usually have one of those modules in your home that will flash light and emit sound in case motion is detected. Of course, you can perfectly connect it to a real siren instead of a buzzer to have a loud sound in case any motion is detected.

To assemble this module, first place the LED in series with the 330 Ohm resistor on the breadboard, with the longest pin of the LED in contact with the resistor. Also place the buzzer on the breadboard.

Then, connect the other side of the resistor to GPIO14 of the Pi and the other part of the LED to one GND pin of the Pi.

For the buzzer, connect the pin marked as **+** on the buzzer to GPIO15 and the other pin of the buzzer to one GND pin of the Pi.

This is the final result:

To configure this module, we will again use the aREST library, so the code will be very similar to the one we used in the previous section:

```
// Modules
var express = require('express');

// Express app
var app = express();

// aREST
var piREST = require('pi-arest')(app);
piREST.set_id('35f5fc');
piREST.set_name('alarm');
piREST.set_mode('bcm');

// Start server
app.listen(3000, function () {
  console.log('Raspberry Pi Zero alarm started!');
});
```

You can now simply grab this code from the GitHub repository of the book or simply paste it into a file called alarm.js using a terminal inside the same folder as the file type:

```
sudo npm install express pi-arest
```

Once the required modules are installed, type the following command to start the project:

```
sudo node alarm.js
```

Finally, let's just try to set the buzzer on; navigate to the IP address of your Pi on port 3000 followed by the digital command on pin 15:

```
http://192.168.0.105:3000/digital/17/1
```

This should immediately set the buzzer on and it should continuously emit sound. To switch it off again, simply type the same command followed by a 0.

Building a wireless security camera

We are now going to build the module that will act as a wireless security camera. You can have one or many of those modules inside your home; it will allow you to observe what is going on in your home from a central location.

The hardware configuration for this part will be really simple, as we are using an USB camera. However, you will need to use an USB hub here, as we will need to connect the USB camera and the usual WiFi dongle on the Raspberry Pi.

This is the final result:

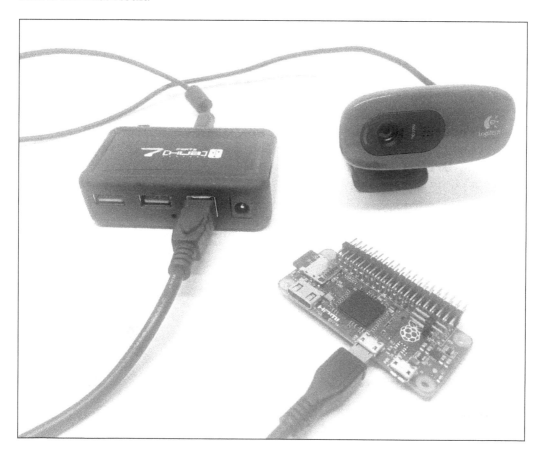

Let's now test the camera first, by taking a simple picture from the command line. You will need to install the fswebcam utility. To do so, simply type the following command inside a terminal:

```
sudo apt-get install fswebcam
```

Then, still from a terminal, you can take a picture with the following command:

```
fswebcam -r 1280x720 image.jpg
```

This will make a lot of messages appear inside the terminal, confirming that the picture has been taken:

```
pi@raspberrypi:~ $ fswebcam -r 1280720 image.jpg
--- Opening /dev/video0...
Trying source module v4l2...
/dev/video0 opened.
No input was specified, using the first.
Adjusting resolution from 1280720x-1 to 1280x960.
--- Capturing frame...
Captured frame in 0.00 seconds.
--- Processing captured image...
Fontconfig warning: ignoring UTF-8: not a valid region tag
Writing JPEG image to 'image.jpg'.
pi@raspberrypi:~ $
```

You can now use an image utility to open the picture you just took. I, for example, used GPicView:

Now that we are sure that the camera is working correctly, we can use it to stream video on the network. For that, we'll use software called MJPG-streamer. To install it, first clone the GitHub repository from a terminal:

```
git clone https://github.com/jacksonliam/mjpg-streamer
```

Then, install some required packages:

```
sudo apt-get install cmake libjpeg62-dev
```

Once that's done, navigate into the folder of the `mjpg-streamer` software and type:

```
sudo make clean all
```

When the compilation of the software is done, type:

```
export LD_LIBRARY_PATH=.
```

Finally, start the software with the following command:

```
./mjpg_streamer -i "./input_uvc.so" -o "./output_http.so -w ./www"
```

You should be able to see a similar output inside the terminal:

```
pi@raspberrypi:~/Work/mjpg-streamer/mjpg-streamer-experimental $ ./mjpg_streamer -i "./input_uvc.so" -o "./output_http.
so -w ./www"
MJPG Streamer Version.: 2.0
 i: Using V4L2 device.: /dev/video0
 i: Desired Resolution: 640 x 480
 i: Frames Per Second.: -1
 i: Format............: JPEG
 i: TV-Norm...........: DEFAULT
UVCIOC_CTRL_ADD - Error at Pan (relative): Inappropriate ioctl for device (25)
UVCIOC_CTRL_ADD - Error at Tilt (relative): Inappropriate ioctl for device (25)
UVCIOC_CTRL_ADD - Error at Pan Reset: Inappropriate ioctl for device (25)
UVCIOC_CTRL_ADD - Error at Tilt Reset: Inappropriate ioctl for device (25)
UVCIOC_CTRL_ADD - Error at Pan/tilt Reset: Inappropriate ioctl for device (25)
UVCIOC_CTRL_ADD - Error at Focus (absolute): Inappropriate ioctl for device (25)
UVCIOC_CTRL_MAP - Error at Pan (relative): Inappropriate ioctl for device (25)
UVCIOC_CTRL_MAP - Error at Tilt (relative): Inappropriate ioctl for device (25)
UVCIOC_CTRL_MAP - Error at Pan Reset: Inappropriate ioctl for device (25)
UVCIOC_CTRL_MAP - Error at Tilt Reset: Inappropriate ioctl for device (25)
UVCIOC_CTRL_MAP - Error at Pan/tilt Reset: Inappropriate ioctl for device (25)
UVCIOC_CTRL_MAP - Error at Focus (absolute): Inappropriate ioctl for device (25)
UVCIOC_CTRL_MAP - Error at LED1 Mode: Inappropriate ioctl for device (25)
UVCIOC_CTRL_MAP - Error at LED1 Frequency: Inappropriate ioctl for device (25)
UVCIOC_CTRL_MAP - Error at Disable video processing: Inappropriate ioctl for device (25)
UVCIOC_CTRL_MAP - Error at Raw bits per pixel: Inappropriate ioctl for device (25)
 o: www-folder-path...: ./www/
 o: HTTP TCP port.....: 8080
 o: username:password.: disabled
 o: commands..........: enabled
```

Then, simply navigate to port 8080 of your Raspberry Pi to access the interface of the streaming software, for example:

```
http://192.168.0.105:8080
```

You should be able to see the interface:

You can now just click on **Stream** to access the live stream from the camera:

From there, you can monitor the live stream from the camera. In the next section, we are going to see how to integrate this stream (and streams from other cameras if you have many) into a central interface.

Note that this project would also work with the official Raspberry Pi camera. However, the first versions of the Raspberry Pi Zero board didn't have the connector for the Raspberry Pi camera, so first make sure if your board has this connector (which is on the side of the board). For more information, visit this link:

https://www.raspberrypi.org/blog/zero-grows-camera-connector/

Creating a security system

In the last section of this chapter, we are going to learn how to integrate all the modules we built in this chapter into a central interface, from which you'll be able to monitor them.

For this project, I ran this last part on my personal computer, but you can, of course, use another Pi Zero board (or any Raspberry Pi board) to run this software.

Let's now see the code for this last section. It will be again composed a main Node.js file for the server, and one HTML and JavaScript files for the interface itself.

Let's first see the Node.js part. It starts by importing all the required modules:

```
// Modules
var express = require('express');
var app = express();
var request = require('request');

// Use public directory
app.use(express.static('public'));
```

Then, you will need to modify the code to put the IP addresses of the Raspberry modules you will be using in the project (except the camera modules, we'll set their IPs directly inside the interface):

```
var motionSensorPi = "192.168.0.104:3000";
var alarmPi = "192.168.0.103:3000"
```

We also define the pins of the different components that are connected to our modules:

```
var buzzerPin = 15;
var ledPin = 14;
var motionSensorPin = 17;
```

Then, we can define the different routes of the project. It starts with a main route that will serve the interface:

```
app.get('/', function (req, res) {

    res.sendfile(__dirname + '/public/interface.html');

});
```

We also need to define a route to get the current state of the alarm:

```
app.get('/alarm', function (req, res) {

    res.json({alarm: alarm});

});
```

Then, we set another route to set the alarm off:

```
app.get('/off', function (req, res) {

    // Set alarm off
    alarm = false;

    // Set LED & buzzer off
    request("http://" + alarmPi + "/digital/" + ledPin + '/0');
    request("http://" + alarmPi + "/digital/" + buzzerPin + '/0');

    // Answer
    res.json({message: "Alarm off"});

});
```

We also start the server itself:

```
var server = app.listen(3000, function() {
    console.log('Listening on port %d', server.address().port);
});
```

Finally, we create a loop that will check the status of the motion sensor every two seconds and set the alarm if motion is detected:

```
setInterval(function() {

    // Get data from motion sensor
    request("http://" + motionSensorPi + "/digital/" +
motionSensorPin,
        function (error, response, body) {

            if (!error && body.return_value == 1) {

                // Activate alarm
                alarm = true;

                // Set LED on
                request("http://" + alarmPi + "/digital/" + ledPin +
'/1');

                // Set buzzer on
                request("http://" + alarmPi + "/digital/" + buzzerPin +
'/1');

            }
        });

}, 2000);
```

Let's now see the interface file, starting by the HTML. It starts by importing all the required libraries and files for the project:

```
<head>
    <script src="https://code.jquery.com/jquery-2.2.4.min.js"></script>
    <link rel="stylesheet" href="https://maxcdn.bootstrapcdn.com/
bootstrap/3.3.6/css/bootstrap.min.css">
    <script src="https://maxcdn.bootstrapcdn.com/bootstrap/3.3.6/js/
bootstrap.min.js"></script>
    <script src="js/script.js"></script>
    <link rel="stylesheet" href="css/style.css">
    <meta name="viewport" content="width=device-width, initial-scale=1">
</head>
```

Then, inside a `<script>` tag on the same page, we'll define some JavaScript functions to integrate the live video stream into the page. We start by declaring the required variables:

```
var imageNr = 0; // Serial number of current image
var finished = new Array(); // References to img objects which have
finished downloading
var paused = false;
```

Then, we declare a function that will create the image layer on the page that will later display the video server:

```
function createImageLayer() {
  var img = new Image();
  img.style.position = "absolute";
  img.style.zIndex = -1;
  img.onload = imageOnload;
  img.onclick = imageOnclick;
  img.src = "http://192.168.0.105:8080/?action=snapshot&n=" +
(++imageNr);
  var webcam = document.getElementById("webcam");
  webcam.insertBefore(img, webcam.firstChild);
}
```

We then define a function to load the next image:

```
function imageOnload() {
  this.style.zIndex = imageNr; // Image finished, bring to front!
  while (1 < finished.length) {
    var del = finished.shift(); // Delete old image(s) from
document
    del.parentNode.removeChild(del);
  }
  finished.push(this);
  if (!paused) createImageLayer();
}
```

We also add the possibility to stop the stream in case we click on the picture:

```
function imageOnclick() { // Clicking on the image will pause the
stream
  paused = !paused;
  if (!paused) createImageLayer();
}
```

We also set the required function on the `<body>` tag, to load the stream when we load the HTML page:

```
<body onload="createImageLayer();">
```

For the interface itself, we first define an indicator for the current status of the alarm:

```
<div class='row voffset50'>

    <div class='col-md-4'></div>
    <div class='col-md-4 text-center'>
      Alarm is <span id='alarm-status'>OFF</span>
    </div>
    <div class='col-md-4'></div>

  </div>
```

In the following section, we create a button to deactivate the alarm if it has been triggered:

```
<div class='row'>

    <div class='col-md-4'></div>
    <div class='col-md-4'>
       <button id='off' class='btn btn-block btn-danger'>Deactivate
Alarm</button>
    </div>
    <div class='col-md-4'></div>

  </div>
```

Finally, we create the element that will hold the live video stream:

```
<div class='row voffset50'>
    <div class='col-md-3'></div>
    <div class='col-md-7'>
      <div id="webcam">
        <noscript>
          <img src="http://192.168.0.105:8080/?action=snapshot" />
        </noscript>
      </div>
    </div>
  </div>
```

Let's now have a look at the JavaScript file. We will first link the button to the correct action on the server:

```
$( "#off" ).click(function() {

    // Deactivate alarm
    $.get('/off');

});
```

For the indicator of the current state of the alarm, we refresh the element of the interface every two seconds:

```
setInterval(function () {

    // Current
    $.get('/alarm', function(data) {

      if (data.alarm == true) {
        $( "#alarm-status" ).text("ON");
      }
      else {
        $( "#alarm-status" ).text("OFF");
      }

    });

}, 2000);
```

It's now finally time to test the last part of the chapter! Grab all the code from the GitHub repository of the book and inside the folder where the code files are type:

```
sudo npm install express request
```

Then, start the application with the following command:

```
sudo node system_interface.js
```

You can now simply navigate to the IP address of the computer or Pi on which the application is running, followed by port 3000. For example:

```
http://192.168.0.100:3000
```

You should immediately see the simple interface that we just created, as well as the live stream from security camera:

You can now try to pass your hand again in front of the motion sensor; you should instantly hear the sound from the buzzer and see the confirmation on your screen inside the interface. Then, simply click on the button to deactivate the alarm.

If you can't see the page, check your firewall: it might be blocking the IP address of your Pi or the port on which the application is running (3000).

Summary

In this chapter, we learned how to build a modular security system based on Raspberry Pi Zero. There are of course many ways to improve this project. For example, you can simply add more modules to the project, like having more motion sensors that triggers the same alarm. You can also use some simple software like Ngrok to access the live video stream remotely, even if you are outside of the WiFi network of your home.

In the next chapter, we are going to dive into the Internet of Things again, and learn how to monitor and control your home from anywhere in the world!

8
Monitor Your Home from the Cloud

In this chapter, we are going to delve more into a very exciting topic related to building a smart home: the Internet of Things. Indeed, today most of the smart homes are connected to the Internet and allows the user the monitor her or his home, even when they are at the other end of the globe.

In this chapter, we are going to learn how to build the three projects that will allow you to monitor your home from a distance. First, we are simply going to add a sensor to our Raspberry Pi Zero and monitor the measurements from a cloud dashboard. After that, we are going to learn how to build our own cloud dashboard to monitor several sensors remotely. Finally, we'll learn how to monitor the live camera stream via a wireless security camera from anywhere in the world. Let's dive in!

Hardware and software requirements

As always, we are going to start with the list of required hardware and software components for the project.

For the sensors, we'll use a simple DHT11 sensor, along with a 4.7k Ohm resistor. We'll also use a PIR motion sensor.

For the camera module, I will use a Logitech C270 webcam. Here, any camera compatible with the UVC protocol would work, which is the case for most of the cameras sold those days.

Finally, you will need the usual breadboard and jumper wires.

This is the list of components that you will need for this whole chapter, not including the Raspberry Pi Zero:

- PIR motion sensor (`https://www.sparkfun.com/products/13285`)
- DHT11 sensor with 4.7k Ohm resistor (`https://www.adafruit.com/products/386`)
- PIR motion sensor (`https://www.adafruit.com/products/189`)
- Logitech C270 USB camera (`http://www.logitech.com/en-us/product/hd-webcam-c270`)
- Breadboard (`https://www.adafruit.com/products/64`)
- Jumper wires (`https://www.adafruit.com/products/1957`)

To connect the camera to your Pi, I also recommend using a USB hub for this chapter, as there is only one USB port on the Pi.

On the software side, you will need to have Node.js installed on your Raspberry Pi Zero board.

Monitoring data from a cloud dashboard

In this first section of the chapter, we are going to connect a temperature and humidity sensor to our Raspberry Pi Zero board and send those measurements to the cloud. Later in this section, we are also going to learn how to visualize those measurements on a dashboard.

We first need to connect the DHT11 sensor to our Pi. First, place the sensor on the board, and then connect the 4.7k Ohm resistor between pin 1 and 2 of the sensor. Then, connect the first pin of the sensor to a 3.3V pin of the Pi, the second pin to GPIO4 of the Raspberry Pi, and finally the last pin of the sensor to a GND pin of the Pi.

The following image is the final result:

We are now going to see how to configure our Raspberry Pi Zero so it automatically sends data to the cloud. For that, we'll use Node.js to send data to a service called **Dweet.io**, which will allow us to easily store data online.

Let's first see the details of the code. First, we declare the modules that we will use for this section:

```
var sensorLib = require('node-dht-sensor');
var request = require('request');
```

After that, we need to give a name to our thing, which is the name we'll use to identify the object storing the measurements on Dweet.io:

```
var thingName = 'mypizero';
```

We will also define a main measurement loop, in which we'll make measurements from the sensor and send those measurements to Dweet.io:

```
var sensor = {
    initialize: function () {
        return sensorLib.initialize(11, 4);
    },
    read: function () {
```

```
        // Readout
        var readout = sensorLib.read();
        console.log('Temperature: ' + readout.temperature.toFixed(2) +
'C, ' +
            'humidity: ' + readout.humidity.toFixed(2) + '%');

        // Log data
        logData(readout);

        // Repeat
        setTimeout(function () {
            sensor.read();
        }, 2000);
    }
};
```

After that, we need to initialize the sensor:

```
if (sensor.initialize()) {
    sensor.read();
} else {
    console.warn('Failed to initialize sensor');
}
```

Let's now see the details of the function that is used to log data on the Dweet.io server:

```
function logData(readout) {

    // Build URL
    var url = "https://dweet.io/dweet/for/" + thingName;
    url += "?temperature=" + readout.temperature.toFixed(2);
    url += "&humidity=" + readout.humidity.toFixed(2);

    // Make request
    request(url, function (error, response, body) {
      if (!error && response.statusCode == 200) {
        console.log(body) // Show response
      }
    });

}
```

We basically form a request to Dweet.io, passing the measurements inside the request URL itself.

It's finally the time to test the project! Grab all the code from the GitHub repository of the book and place it inside a folder on your Pi. Then, inside this folder, type the following command with a terminal:

```
npm install node-dht-sensor
```

This will install the required sensor library. Then, install the `request` module with the following command:

```
sudo npm install request
```

Finally, you can start the software by typing:

```
sudo node sensor_cloud_log.js
```

You should immediately see the answer from Dweet.io as the data is recorded to the cloud:

```
pi@raspberrypi:~/Work/rpi-book/smart-homes-pi-zero/08 $ sudo node sensor_cloud_log.js
Temperature: 27.00C, humidity: 32.00%
{"this":"succeeded","by":"dweeting","the":"dweet","with":{"thing":"mypizero","created":"2016-08-21T
08:10:56.969Z","content":{"temperature":27,"humidity":32},"transaction":"609ff155-5f77-40c1-a856-f8
fe96dfa686"}}
Temperature: 27.00C, humidity: 32.00%
{"this":"succeeded","by":"dweeting","the":"dweet","with":{"thing":"mypizero","created":"2016-08-21T
08:10:58.351Z","content":{"temperature":27,"humidity":32},"transaction":"15bd590a-c669-4ccd-88db-c4
a647033ab5"}}
Temperature: 27.00C, humidity: 32.00%
{"this":"succeeded","by":"dweeting","the":"dweet","with":{"thing":"mypizero","created":"2016-08-21T
08:11:01.398Z","content":{"temperature":27,"humidity":32},"transaction":"4f885179-c7f2-4b26-a6ce-48
9a91532fdc"}}
Temperature: 27.00C, humidity: 32.00%
{"this":"succeeded","by":"dweeting","the":"dweet","with":{"thing":"mypizero","created":"2016-08-21T
08:11:04.911Z","content":{"temperature":27,"humidity":32},"transaction":"0261a12a-c071-45a6-ad57-9f
b825de1f36"}}
```

You can actually already visualize this data right in your web browser, by typing the following URL:

```
←    C    🔒 https://dweet.io/get/latest/dweet/for/mypizero          ☆  ...

{"this":"succeeded","by":"getting","the":"dweets","with":[{"thing":"mypizero","created":"2016-
08-21T08:11:12.755Z","content":{"temperature":27,"humidity":32}}]}
```

This is nice, but it's not great to actually visualize data as it is recorded. That's why we are now going to use Freeboard.io, which is a service that will allow us to create cloud dashboards using the Dweet.io data.

You can already create an account at:

```
http://freeboard.io/
```

Inside Freeboard.io, first create a new dashboard:

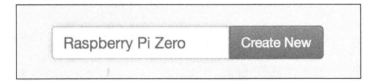

Then, add a new datasource with the following parameters:

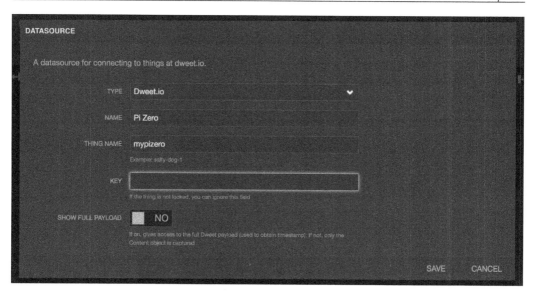

This will basically link your dashboard to the 'thing' that is storing your data on Dweet.io. After that, you'll see that the connection is active inside the dashboard itself:

Now, create a new pane inside your dashboard and also a new **Gauge** widget for the temperature, using the following parameters:

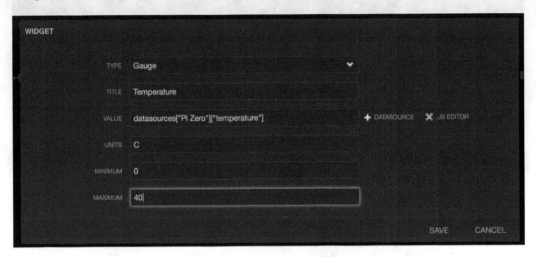

You should be able to immediately see the temperature measurements being displayed in the dashboard:

Now, do the same for humidity, using the following parameters:

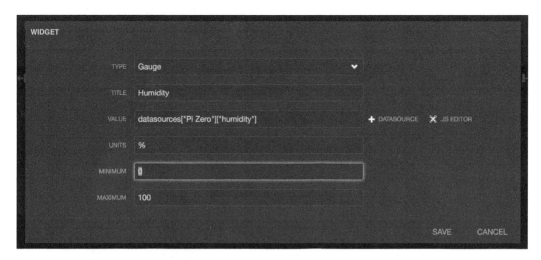

You should now have both gauges inside your dashboard, giving you an immediate glance at the temperature and humidity inside your home:

You can now add more visualizations of the same data. For example, I added two additional widgets of the type **sparkline** for each measurement, giving me an instant view of the recent history of each variable:

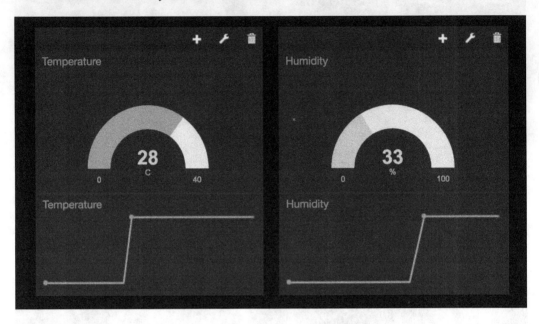

Creating a cloud dashboard for your devices

In the second part of this chapter, we are going to add a motion sensor to the project we built in the first part and also learn how to monitor all those sensors from a single dashboard. I will connect all the sensors to a single Raspberry Pi Zero board, but you could of course have them connected to several boards that are in different parts of your smart home.

The project itself will be really easy to assemble. First, make sure that you followed all the instructions from the previous project. Then, simply connect the motion sensor to the project: VCC goes to the 3.3V pin of the Raspberry Pi, GND to GND, and the SIG pin of the sensor is connected to Raspberry Pi GPIO18.

The following image is the final result:

Let's now see how to configure the project. In order to access the measurements from anywhere in the world, we'll use the aREST framework again, which we have already used in several projects of the book. However, here we'll use the cloud access of aREST that will allow us to access those measurements from anywhere.

Inside the code itself, we first include all the required modules:

```
var sensorLib = require('node-dht-sensor');
var express = require('express');
var app = express();
var piREST = require('pi-arest')(app);
```

We then define the ID and the name of the board:

```
piREST.set_id('73gutg');
piREST.set_name('pi_zero_cloud');
piREST.set_mode('bcm');
```

Note that as the ID is unique for each Raspberry Pi board, you need to change it and insert your own ID here. Then, inside the main measurement loop, we expose the temperature and humidity measurements to the aREST framework:

```
piREST.variable('temperature', readout.temperature.toFixed(2));
piREST.variable('humidity', readout.humidity.toFixed(2));
```

We also do the same for a variable called `motion` that depends on the current state of the motion sensor:

```
piREST.digitalRead(18, function(data) {

        if (data == 1) {
            piREST.variable('motion', "Motion Detected");
        }
        else {
            piREST.variable('motion', "No Motion");
        }

    });
```

After that, we connect to the aREST cloud server:

```
piREST.connect();
```

Finally, we start the server with the following code:

```
var server = app.listen(80, function() {
    console.log('Listening on port %d', server.address().port);
});
```

It's now finally time to test the project! Inside the same folder as you put the files of the first project of this chapter, type:

sudo npm install express pi-arest

When the modules are installed, start the project using:

sudo node sensor_cloud_arest.js

This will immediately make the project connect to the aREST.io cloud server. You can actually test it by typing the following URL inside your favorite web browser, of course by changing the ID of your device:

```
←  →  C    🔒 https://cloud.arest.io/73gutg/temperature

{"id":"73gutg","name":"pi_zero_cloud","hardware":"rpi","connected":true,"temperature":"27.00"}
```

In order to display this data inside a dashboard, we are going to use the dashboard of the aREST framework that you can access from:

```
http://dashboard.arest.io/
```

From there, create a new account and then a new dashboard:

Inside this dashboard, add a new element with the following parameters, which will be used to display the temperature inside the dashboard:

Once it is done, you can do the same for humidity. You should now have both the data showing up inside your dashboard:

Finally, do a similar operation with the motion variable:

You should now have all the variables measured by your Pi displayed inside the same dashboard:

Congratulations, you can now monitor your home from anywhere in the world! Also note that you can have measurements from several Raspberry Pi boards displayed in the same cloud dashboard.

Accessing your security camera from anywhere

Inside the last section of this chapter, we are going to revisit a project we built in the last chapter: the wireless security camera. Here, we are going to learn how to visualize the live video stream coming from a camera from anywhere in the world.

Assembling this project is really simple; you simply need to plug the USB camera to the Raspberry Pi using the USB hub as you also need to plug the Wi-Fi dongle. This is the final result:

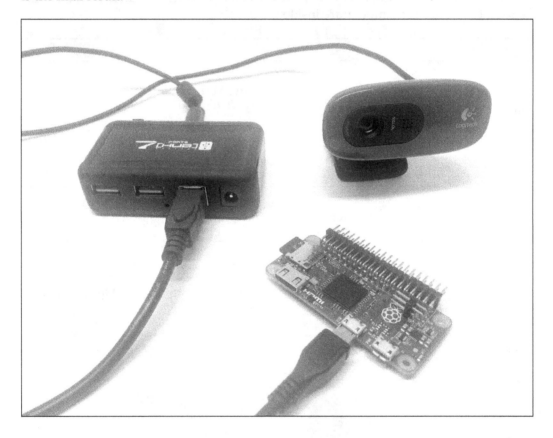

Now, we are again going to use the MJPEG-streamer software to create a live video stream from the camera. If you haven't installed it yet, you can check the last chapter of the book to learn how to install it.

Then, go inside the folder where the software is installed, and type:

```
./mjpg_streamer -i "./input_uvc.so" -o "./output_http.so" -w ./www"
```

This will immediately start the streaming software on your Raspberry Pi:

```
pi@raspberrypi:~/Work/mjpg-streamer/mjpg-streamer-experimental $ ./mjpg_streamer -i "./input_uvc.so" -o "./ou
tput_http.so -w ./www"
MJPG Streamer Version.: 2.0
 i: Using V4L2 device.: /dev/video0
 i: Desired Resolution: 640 x 480
 i: Frames Per Second.: -1
 i: Format............: JPEG
 i: TV-Norm...........: DEFAULT
UVCIOC_CTRL_ADD - Error at Pan (relative): Inappropriate ioctl for device (25)
UVCIOC_CTRL_ADD - Error at Tilt (relative): Inappropriate ioctl for device (25)
UVCIOC_CTRL_ADD - Error at Pan Reset: Inappropriate ioctl for device (25)
UVCIOC_CTRL_ADD - Error at Tilt Reset: Inappropriate ioctl for device (25)
UVCIOC_CTRL_ADD - Error at Pan/tilt Reset: Inappropriate ioctl for device (25)
UVCIOC_CTRL_ADD - Error at Focus (absolute): Inappropriate ioctl for device (25)
UVCIOC_CTRL_MAP - Error at Pan (relative): Inappropriate ioctl for device (25)
UVCIOC_CTRL_MAP - Error at Tilt (relative): Inappropriate ioctl for device (25)
UVCIOC_CTRL_MAP - Error at Pan Reset: Inappropriate ioctl for device (25)
UVCIOC_CTRL_MAP - Error at Tilt Reset: Inappropriate ioctl for device (25)
UVCIOC_CTRL_MAP - Error at Pan/tilt Reset: Inappropriate ioctl for device (25)
UVCIOC_CTRL_MAP - Error at Focus (absolute): Inappropriate ioctl for device (25)
UVCIOC_CTRL_MAP - Error at LED1 Mode: Inappropriate ioctl for device (25)
UVCIOC_CTRL_MAP - Error at LED1 Frequency: Inappropriate ioctl for device (25)
UVCIOC_CTRL_MAP - Error at Disable video processing: Inappropriate ioctl for device (25)
UVCIOC_CTRL_MAP - Error at Raw bits per pixel: Inappropriate ioctl for device (25)
 o: www-folder-path...: ./www/
 o: HTTP TCP port.....: 8080
 o: username:password.: disabled
 o: commands..........: enabled
```

Now, I want you to open another terminal window or tab, as we will need to run another software while the streaming software is running. We are going to use software called **Ngrok**; this will allow us to access the video stream from anywhere in the world.

In the second terminal window, type:

```
wget https://bin.equinox.io/c/4VmDzA7iaHb/ngrok-stable-linux-arm.zip
```

This will download Ngrok on your computer. Then, unzip the file with:

```
unzip ngrok-stable-linux-arm.zip
```

Finally, start Ngrok using the following command:

```
./ngrok 8080
```

This will basically create a web URL that you can use to access your Pi on port `8080`, which is precisely the port on which the streaming software is running. Once Ngrok is running, you should be able to see the URL you need inside a window:

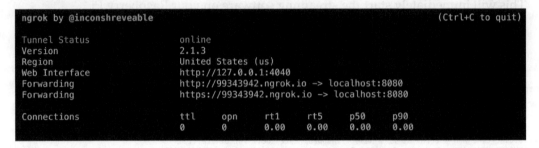

You can now simply copy this URL and type it inside any web browser. You should then immediately be able to see the interface created by the streaming software:

Now, simply click on **Stream** to see the live stream coming from the board:

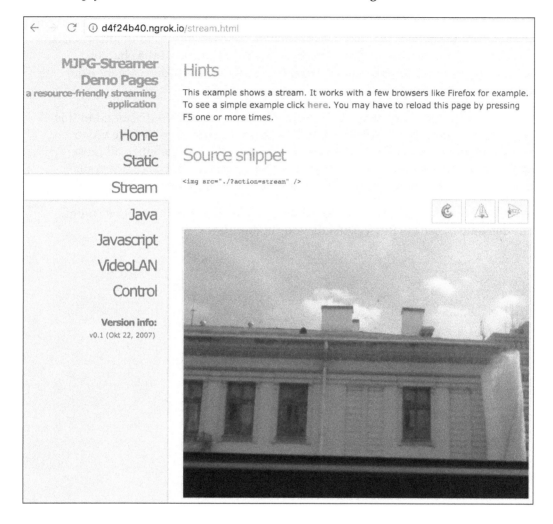

Congratulations, you can now access your wireless security camera from anywhere in the world and monitor your home remotely!

Summary

In this chapter, we learned how to use the IoT to monitor our homes remotely from anywhere in the world. We first learned how to log data in the cloud and visualize this data using two different IoT platforms. Then, at the end of the chapter, we learned how to visualize the live stream coming from a video camera from anywhere in the world.

There are of course many ways to improve the projects that we discussed in this chapter. You could, for example, create a web server that takes the live video streams from several video cameras and then use Ngrok to visualize all those live video streams at once. This will instantly give you a video security system, that you can use to monitor your home from anywhere in the world!

In the next chapter, we are going to continue diving into the Internet of Things, but this time to control devices inside your home.

9
Control Appliances from Anywhere

In this chapter, we are going to continue exploring the IoT field, and learn how we can use it for a smart home. In the previous chapter, we learned how to make our home send data to the cloud, using the Raspberry Pi Zero board. Here, we'll actually do the opposite; we are going to learn how to control appliances in your home from anywhere in the world.

We are going to start with a simple example, controlling a simple LED from anywhere in the world. Then, we'll see how to control lamps using the same principles. After that, we are going to use IFTTT again (as we did in *Chapter 6, Sending Notifications using Raspberry Pi Zero*) to build two exciting applications: a lamp that switches on when motion is detected and a cloud thermostat. Let's start!

Hardware and software requirements

As always, we are going to start with a list of required hardware and software components for the project.

For the devices to control, we'll use a simple LED with a 330 Ohm resistor and then the PowerSwitch Tail Kit that we already used in several chapters of this book.

For the sensors, which we will need at the end of the chapter, we'll use a simple DHT11 sensor, along with a 4.7k Ohm resistor. We'll also use a PIR motion sensor.

Finally, you will need the usual breadboard and jumper wires.

The following is the list of components that you will need for this whole chapter, not including the Raspberry Pi Zero:

- LED (https://www.sparkfun.com/products/9590)
- 330 Ohm resistor (https://www.sparkfun.com/products/11507)
- PowerSwitch Tail Kit (https://www.adafruit.com/products/268)
- PIR motion sensor (https://www.sparkfun.com/products/13285)
- DHT11 sensor with 4.7k Ohm resistor (https://www.adafruit.com/products/386)
- PIR motion sensor (https://www.adafruit.com/products/189)
- Breadboard (https://www.adafruit.com/products/64)
- Jumper wires (https://www.adafruit.com/products/1957)

On the software side, you will just need to have Node.js installed on your Raspberry Pi Zero board.

Control a LED from anywhere in the world

For the first project of this chapter, we are simply going to learn how to control a simple LED from a cloud dashboard.

For this project, you will need a LED and a 330 Ohm resistor. For the connection of the components to the Pi, you can refer to *Chapter 4, Control Appliances from the Raspberry Pi Zero* of this book, in which you will learn how to connect those components. You need to connect the LED to GPIO14 of the Pi.

This is the final result:

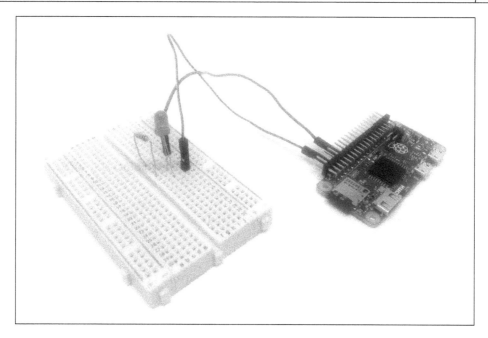

Let's now see how to configure the board, so we can control it from the cloud. To do so, we'll use the aREST framework that we have already used several times in this book. The following is the complete code for this part:

```
// Required modules
var express = require('express');
var app = express();
var piREST = require('pi-arest')(app);

// Thing name
piREST.set_id('98t52d');
piREST.set_name('pi_zero_cloud');
piREST.set_mode('bcm');

// Connect to cloud.aREST.io
piREST.connect();

// Start server
var server = app.listen(80, function() {
    console.log('Listening on port %d', server.address().port);
});
```

Of course, make sure that you modify the ID of the board inside the code, as it will identify your board on the aREST cloud server.

Then, either put this code inside a file or get the whole code from the GitHub repository of the book.

Then, inside a terminal, type:

```
sudo npm install pi-arest express
```

This will install the required modules for this section. Then, start the software using:

```
sudo node arest_control.js
```

You can then go to the following website and register:

```
http://dashboard.arest.io/
```

We'll basically use this site to build a cloud dashboard to control our LED. Once you create an account, create a new dashboard:

Inside this dashboard, create a new element by giving the ID of your Raspberry Pi that you set in the code earlier. I chose a 'Push' button as the type of the element and 14 as the pin to control.

Once you have created the element, this is what you should be able to see on the dashboard:

You can now test the push button of the dashboard: when you keep the button pressed ON, you should immediately be able to see the LED turning ON on your Pi. Note that this is purely done in the cloud, so you can now control your LED from anywhere in the world!

Creating several lamps from the cloud

In the second project of this chapter, we are going to apply what we learned earlier and control several lamps from the cloud using a single dashboard.

To actually assemble the project, I recommend checking *Chapter 4, Control Appliances from the Raspberry Pi Zero* where we saw how to connect the PowerSwitch Tail Kit to the Raspberry Pi Zero. You need to connect the PowerSwitch to GPIO14 of the Raspberry Pi board.

This is how one module looks like:

Now, configure each board with the exact same code as in the previous section and give a different ID to each board. I also recommend changing the name of the boards inside the code; for example, to know where you placed them in your home (bedroom, living room, and so on).

Then, go back to the website where we created a cloud dashboard and create a new dashboard:

In there, create a new On/Off element for each lamp you want to control, on pin **14**:

This is how your dashboard should look like at the end:

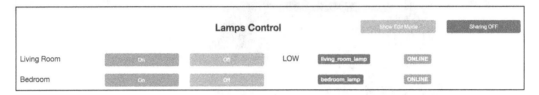

You can now try it: whenever you click on one of the **On** buttons, the lamp connected to this Raspberry Pi should immediately turn on.

Make a motion-activated lamp using IFTTT

In this section, we are now going to use what we learned in this chapter and combine it with what we already learned in the previous chapters about the web service IFTTT. We are going to use this knowledge to build a lamp that is automatically activated when motion is detected by a motion sensor.

For this section, you will need two Raspberry Pi modules: one with a motion sensor and one connected to a lamp via the PowerSwitch tail. To learn how to assemble these modules, please refer to the previous chapters of the book.

This is the assembled Raspberry Pi Zero with a PIR motion sensor on GPIO18:

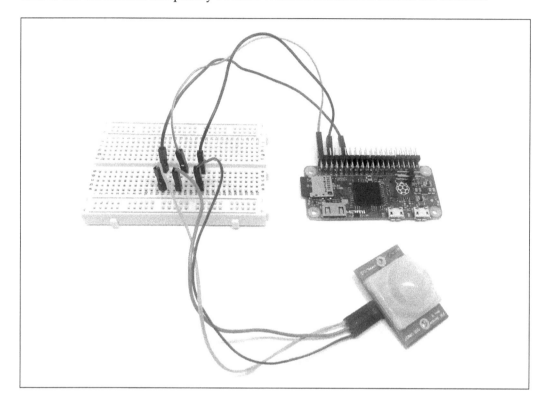

We are first going to create the IFTTT recipes so you and the two boards can communicate. First, make sure that the Maker channel is activated on your IFTTT account:

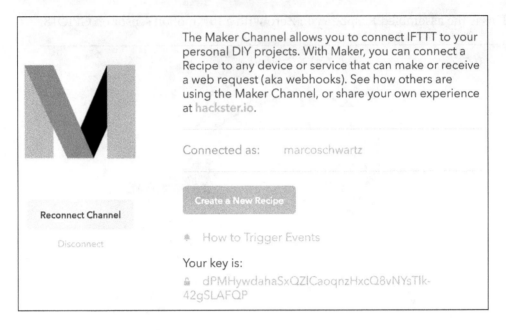

The Maker Channel allows you to connect IFTTT to your personal DIY projects. With Maker, you can connect a Recipe to any device or service that can make or receive a web request (aka webhooks). See how others are using the Maker Channel, or share your own experience at hackster.io.

Connected as: marcoschwartz

Reconnect Channel

Disconnect

Create a New Recipe

⚡ How to Trigger Events

Your key is:

🔒 dPMHywdahaSxQZICaoqnzHxcQ8vNYsTlk-42gSLAFQP

Then, create a new recipe, with the Maker channel as the trigger:

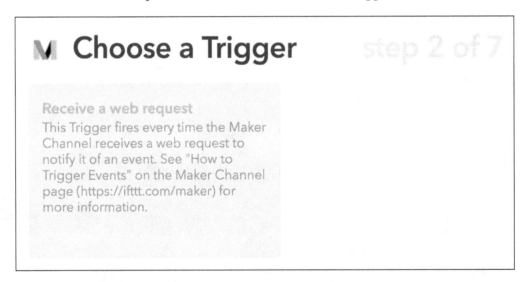

Ⅿ **Choose a Trigger**   step 2 of 7

Receive a web request
This Trigger fires every time the Maker Channel receives a web request to notify it of an event. See "How to Trigger Events" on the Maker Channel page (https://ifttt.com/maker) for more information.

For the event, enter `motion_detected`:

M Complete Trigger Fields

Receive a web request

M Event Name

motion_detected

The name of the event, like "button_pressed" or "front_door_opened"

Create Trigger

Choose the Maker channel as the action channel:

For the action itself, choose **Make a web request**:

M Choose an Action

Make a web request
This Action will make a web request to
a publicly accessible URL. NOTE:
Requests may be rate limited.

We want to activate the lamp if motion is detected, enter the following URL as the action when this recipe is activated:

You can now save this recipe. Of course, we also want to switch the light off again when no motion is detected, so we need to create another recipe with this event:

This action is the same as the previous one, but with a 0 at the end, meaning we are switching the light off:

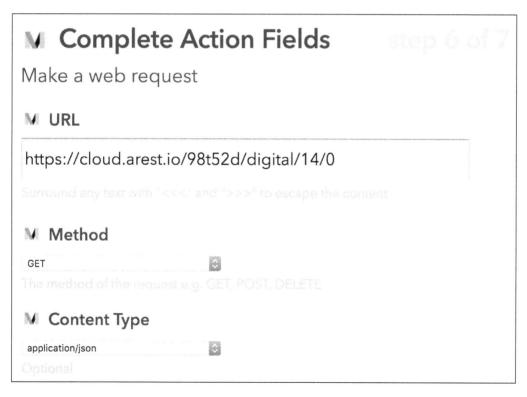

At the end, you should have both recipes active inside the dashboard:

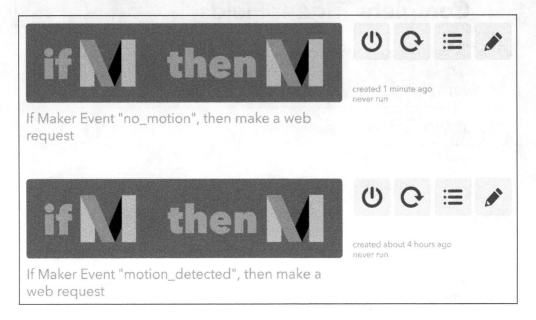

If Maker Event "no_motion", then make a web request

If Maker Event "motion_detected", then make a web request

Let's now see how to configure the boards. For the board connected to the PowerSwitch tail, you can use the same code that we used in the previous sections of this chapter.

For the motion sensor board, the code starts by including the required modules:

```
var request = require('request');
var gpio = require('rpi-gpio');
```

Then, we define the IFTTT, as well as the name of the two events we used in recipes:

```
var key = "key"
var eventOnName = 'motion_detected';
var eventOffName = 'no_motion';
```

After that, we define the pin on which the sensor is connected:

```
var motionSensorPin = 18;
var motionSensorState = false;
```

We create a main measurement loop in which we check the status of the sensor every second:

```
setInterval(function() {

    // Check sensor
```

```
    gpio.setup(motionSensorPin, gpio.DIR_IN, checkSensor);

}, 1000);
```

If the state of the sensor changes, we send the right command to IFTTT:

```
function checkSensor() {
  gpio.read(motionSensorPin, function(err, value) {

      // If motion is detected
      if (value == true && motionSensorState == false) {

        // Send event
        alertIFTTT(eventOnName);

      }

      // No motion anymore
      if (value == false && motionSensorState == true) {

        // Send event
        alertIFTTT(eventOffName);

      }

      // Set status
      motionSensorState = value;
  });
}
```

Here is the detail of the function that sends the alert to IFTTT:

```
function alertIFTTT(eventName) {

  // Send alert to IFTTT
  console.log("Sending alert to IFTTT");
  var url = 'https://maker.ifttt.com/trigger/' + eventName + '/with/
key/' + key;
  request(url, function (error, response, body) {
    if (!error && response.statusCode == 200) {
      console.log("Alert sent to IFTTT");
    }
  });
}
```

You can now grab all the code from the GitHub repository of the book and extract it in a folder on the Pi with the motion sensor. Then, inside a terminal, install the required modules with:

```
sudo npm install request rpi-gpio
```

Once that's done, start the software with:

```
sudo node motion_trigger.js
```

Also make sure that the aREST sketch is still running the other board. Then, simply pass your hand in front of the sensor; it should immediately light up the lamp that is connected to the other Raspberry Pi board.

Build an automated cloud thermostat

In the last section of this chapter, we are going to apply what we learned in the previous section, but this time to build a cloud thermostat that will work using IFTTT.

Apart from the Raspberry Pi Zero that will control an electrical heater via the PowerSwitch Tail, you will need another Raspberry Pi Zero with a DHT11 sensor that we have already used several times in this book. In order to assemble this module, I recommend checking for example the second chapter of this book.

Once you have your two modules assembled, go again to IFTTT and create a new recipe, using the Maker channel for the trigger and for the action channels.

For the trigger, enter the following event:

Receive a web request

This Trigger fires every time the Maker Channel receives a web request to notify it of an event. See "How to Trigger Events" on the Maker Channel page (https://ifttt.com/maker) for more information.

Ｍ **Event Name**

temperature_low

The name of the event, like "button_pressed" or "front_door_opened"

Of course, if the temperature is too low, it means that we want to activate the heater. We therefore need to send this command to the board that controls the heater:

Once this recipe is created, create another for the `temperature_high` event:

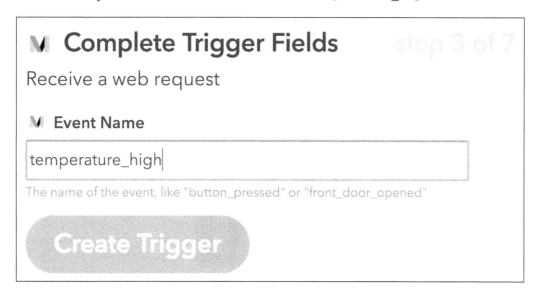

When the temperature is too high, we automatically switch off the heater:

At the end, you should have two new recipes in your dashboard:

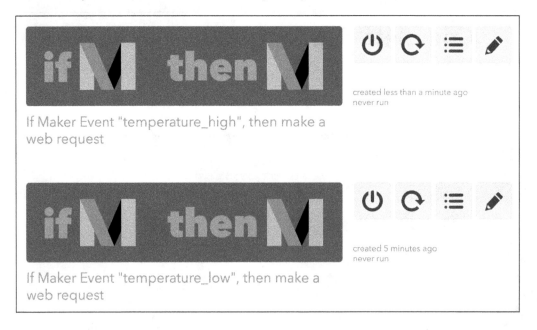

For the board that controls the electrical heater, just use the same software as before.

For the board with the DHT11 sensor, we first need to include the required modules:

```
var request = require('request');
var sensorLib = require('node-dht-sensor');
```

Then, we define the key for the Maker channel and the name of the two events:

```
var key = "key";
var eventNameLow = 'temperature_low';
var eventNameHigh = 'temperature_high';
```

We also declare on which pin the DHT11 sensor is connected to:

```
var sensorPin = 18;
```

Then, we set a target (the temperature we want to reach) and a tolerance:

```
var target = 25;
var tolerance = 1;
```

After that, we create the main measurement loop, in which we check for the right event to send to IFTTT:

```
var sensor = {
    initialize: function () {
        return sensorLib.initialize(11, sensorPin);
    },
    read: function () {

        // Read
        var readout = sensorLib.read();
        temperature = readout.temperature.toFixed(2);
        console.log('Current temperature: ' + temperature);

          if (temperature < target - tolerance) {

            // Send event
            alertIFTTT(temperature_low);

          }

          if (temperature > target + tolerance) {

            // Send event
            alertIFTTT(temperature_high);
```

```
        }

            // Repeat
            setTimeout(function () {
                sensor.read();
            }, 2000);
        }
    };
```

We also end the sketch by initializing the sensor:

```
    // Init sensor
    if (sensor.initialize()) {
        sensor.read();
    } else {
        console.warn('Failed to initialize sensor');
    }
```

Here are the details of the function that we use to make a request:

```
    // Make request
    function alertIFTTT(eventName) {

        // Send alert to IFTTT
        console.log("Sending alert to IFTTT");
        var url = 'https://maker.ifttt.com/trigger/' + eventName + '/with/
    key/' + key;
        request(url, function (error, response, body) {
            if (!error && response.statusCode == 200) {
                console.log("Alert sent to IFTTT");
            }
        });
    }
```

It's now time to test the thermostat project! Simply get all the code from the GitHub repository of the project and then type:

sudo npm install node-dht-sensor request

Once that's done, start the software with:

sudo node temperature_trigger.js

You should now be able to see that the Raspberry Pi, with the electrical heater control, will automatically react based on the temperature, even if both boards are not in the same Wi-Fi network!

Summary

In this chapter, we continued to explore the field of the Internet of Things and we used it to control our smart home remotely. We first saw how to control devices remotely, like simple LEDs and lamps. Then, we learned how to combine what we learned with IFTTT to create more complex projects, for example making a cloud thermostat.

You can of course improve what you learned in this chapter. For example, you can combine what you learned in this chapter and the previous chapter to create cloud dashboards from where you can both monitor your devices and control other devices remotely.

In the next chapter, we are going to use everything that we learned in this book to create a complete home automation system based on the Raspberry Pi Zero.

10
Building a Home Automation System with Raspberry Pi Zero Boards

As this is the final chapter of this book, we are going to integrate everything that we learned in the book to build a complete home automation system based on the Raspberry Pi Zero. We are first going to see how to assemble and configure several modules based on the Raspberry Pi Zero boards, and then learn how to create a server that will communicate with the boards. Note that this server will be able to run on your own computer, and also on a Raspberry Pi Zero board.

We will then learn how to define behaviors inside the code, for example to send you an alert if motion is detected. After that, we are going to learn how to create advanced behavior inside the server, and finally, we'll see how to access your home automation system from anywhere in the world. Let's start!

Hardware and software requirements

As always, we are going to start with a list of required hardware and software components for the project. We are going to use four modules in total: a sensor module, an appliance control module, a motion sensor module, and a camera module.

Of course, you will need one Raspberry Pi Zero board for each module you use, along with supporting components, such as SD cards and power supplies. If you plan to run the central interface on a Raspberry Pi Zero as well, you will need an additional Pi Zero board.

For the sensors module, we'll use a simple DHT11 sensor, along with a 4.7k Ohm resistor.

For the appliance control module, we'll use the PowerSwitch Tail Kit that we already used in several chapters of this book.

For the motion sensor module, we'll use a simple PIR motion sensor.

For the camera module, we are going to use the C270 HD camera from Logitech. However, you can use any USB camera here.

You will also need the usual breadboard and jumper wires.

This is the list of components that you will need for this whole chapter, not including the Raspberry Pi Zero:

- LED (`https://www.sparkfun.com/products/9590`)
- 330 Ohm resistor (`https://www.sparkfun.com/products/11507`)
- PowerSwitch Tail Kit (`https://www.adafruit.com/products/268`)
- PIR motion sensor (`https://www.sparkfun.com/products/13285`)
- DHT11 sensor with 4.7k Ohm resistor (`https://www.adafruit.com/products/386`)
- PIR motion sensor (`https://www.adafruit.com/products/189`)
- Logitech C270 camera (`http://www.logitech.com/en-us/product/hd-webcam-c270`)
- Breadboard (`https://www.adafruit.com/products/64`)
- Jumper wires (`https://www.adafruit.com/products/1957`)

On the software side, you just need to have Node.js installed on your Raspberry Pi Zero boards.

Building all the modules

In the first part of this chapter, we are going to see how to build the modules that we will use in our home automation system. As we have already seen how to build all these modules in the book, I will simply point to the correct chapters to build all the modules.

The first module you need to build is the sensor module, which is with the DHT11 sensor. To learn how to build this module, please refer to *Chapter 2, Measure Data Using Your Raspberry Pi Zero Board*, of the book.

This is what you should get at the end:

For the module that will be used to control appliances in your home, such as lamps, please refer to *Chapter 3, Building a Smart Home Thermostat,* to know how to assemble the module.

This is what you should get at the end:

For the motion sensor module, which is basically composed of a PIR motion sensor connected to the Pi, you can refer to *Chapter 6, Sending Notifications using Raspberry Pi Zero*. You will get the following outcome at the end:

Finally, for the camera module, simply connect the USB camera to the Raspberry Pi zero board using a USB hub. This is the result:

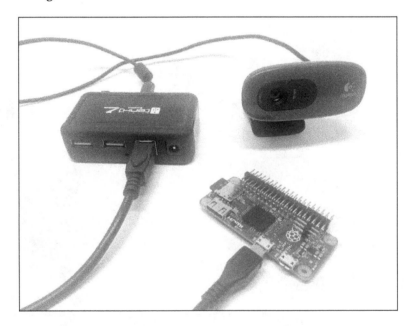

Once you have all those modules, connect them to a source of power and make sure that the latest version of Raspbian is installed on them (along with Node.js), and also make sure they are accessible though Wi-Fi.

Configuring the modules

We are now going to configure each of the modules of our home automation system, so we can access them remotely later. The goal here is to configure the modules to respond to commands coming from our central server and to not act as independent units, as shown in earlier chapters.

Let's start with the sensors module; this is the complete code for this module:

```
// Modules
var express = require('express');
var sensorLib = require('node-dht-sensor');

// Express app
var app = express();

// aREST
var piREST = require('pi-arest')(app);
piREST.set_id('4g0d7f');
piREST.set_name('sensor_module');
piREST.set_mode('bcm');

// Start server
app.listen(3000, function () {
  console.log('Raspberry Pi Zero motion sensor started!');
});

// Sensor loop
var sensor = {
    initialize: function () {
        return sensorLib.initialize(11, 4);
    },
    read: function () {
        var readout = sensorLib.read();
        console.log('Temperature: ' + readout.temperature.toFixed(2) +
'C, ' +
            'humidity: ' + readout.humidity.toFixed(2) + '%');
        setTimeout(function () {
            sensor.read();
        }, 2000);
```

```
    }
  };

  if (sensor.initialize()) {
      sensor.read();
  } else {
      console.warn('Failed to initialize sensor');
  }
```

As you can see, we are once again using the aREST framework to get access to our Raspberry Pi Zero data. Here, we expose two variables to the API that contain the measurements done by the sensor.

You can now grab this code from the GitHub repository of the book or simply grab the code and paste it inside a file.

Then, using a terminal from the folder where the code files are, type the following:

```
npm install node-dht-sensor
```

Once this module is installed, install the rest of the modules with the following command:

```
sudo npm install pi-arest
```

Now, start the project with this command:

```
sudo node sensor_module.js
```

Note that you will need the IP addresses of all the Raspberry Pi later, so it is a good time to actually check the IP address of each Raspberry Pi board you are configuring. To get the IP address of a Pi, simply type the following command inside a terminal:

```
ifconfig
```

Let's now configure the module that will be used to control the appliances remotely. Here is the complete code for this part:

```
// Modules
var express = require('express');

// Express app
var app = express();

// Use public directory
app.use(express.static('public'));

// aREST
```

```
var piREST = require('pi-arest')(app);
piREST.set_id('34f5eQ');
piREST.set_name('lamp_module');
piREST.set_mode('bcm');

// Start server
app.listen(3000, function () {
  console.log('Raspberry Pi Zero lamp module started!');
});
```

You basically just need to change the name of the module inside the code if you wish. Then, grab the code from the GitHub repository of the book and type the following command:

```
sudo npm install pi-arest express
```

After that, start the code with this command:

```
sudo node lamp_module.js
```

We are now going to see how to configure the module connected to the PIR motion sensor. For this, here is the complete code:

```
// Modules
var express = require('express');

// Express app
var app = express();

// aREST
var piREST = require('pi-arest')(app);
piREST.set_id('47g40f');
piREST.set_name('motion_module');
piREST.set_mode('bcm');

// Start server
app.listen(3000, function () {
  console.log('Raspberry Pi Zero motion sensor started!');
});
```

Again, grab the code from the GitHub repository of the book and type this command:

```
sudo npm install pi-arest express
```

Then, start the software with this command:

```
sudo node motion_module.js
```

Finally, let's see how to configure the module with the USB camera. You first need to clone the following GitHub repository using a terminal:

```
git clone https://github.com/jacksonliam/mjpg-streamer
```

Then, install the required packages:

```
sudo apt-get install cmake libjpeg62-dev
```

Once that's done, navigate to the `mjpg-streamer` software folder and type this:

```
sudo make clean all
```

When the compilation of the software is done, type the following:

```
export LD_LIBRARY_PATH=.
```

Finally, start the software with the following command:

```
./mjpg_streamer -i "./input_uvc.so" -o "./output_http.so -w ./www"
```

You should be able to see a lot of text output inside the terminal, meaning that the streaming software is active. Make sure to not stop this software, as we'll need to access the stream remotely from our central server.

Integrating the modules into a single interface

Now that our modules are up and running, we are going to learn how to integrate everything into a single interface, so you will be able to run it on your computer or on another Raspberry Pi. You will then be able to control and monitor your smart home from a single interface.

We will first configure the server that will allow us to connect all the modules that we configured earlier. Then, we'll build an interface on top of that.

The code for the server starts by importing the required modules:

```
// Modules
var express = require('express');
var request = require('request');

// Express app
var app = express();
```

After that, this is where we'll define the IP addresses of the different modules in our home automation system:

```
// Raspberry Pi boards IP addresses
var motionSensorPi = "192.168.0.101:3000";
var sensorPi = "192.168.0.102:3000"
var lampPi = "192.168.0.103:3000"
```

We also need to define the pins on which the lamp and the motion sensor are connected:

```
// Pins
var lampPin = 12;
var motionSensorPin = 17;
```

Of course, if you connect the components to different pins, you will need to change that here.

Then, we can declare the folder in which we will store the files for the interface, as well as the main route of the application that will serve the interface:

```
// Use public directory
app.use(express.static('public'));

// Routes
app.get('/', function (req, res) {

  res.sendfile(__dirname + '/public/interface.html');

});
```

We then create a route that will send us back the state of the motion sensor, by calling the required command on the motion sensor module:

```
app.get('/motion', function (req, res) {

  request("http://" + motionSensorPi + "/digital/" + motionSensorPin,
    function (error, response, body) {

      // Answer
      answer = {
        status: body.return_value
      };
      res.json(answer);

    });

});
```

We also define a route that will give us the temperature measured by the DHT11 sensor, by calling the corresponding Raspberry Pi:

```
app.get('/temperature', function (req, res) {

  request("http://" + sensorPi + "/temperature",
    function (error, response, body) {

      // Answer
      answer = {
        temperature: body.temperature
      };
      res.json(answer);

  });

});
```

We do the same for humidity:

```
app.get('/humidity', function (req, res) {

  request("http://" + sensorPi + "/humidity",
    function (error, response, body) {

      // Answer
      answer = {
        humidity: body.humidity
      };
      res.json(answer);

  });

});
```

Finally, we define a route to turn on the module that control appliances:

```
app.get('/on', function (req, res) {

  request("http://" + lampPi + "/digital/" + lampPin + '/1');

  // Answer
  answer = {
```

```
    status: 1
  };
  res.json(answer);

});
```

We also define a similar route to turn the appliance off again:

```
app.get('/off', function (req, res) {

  request("http://" + lampPi + "/digital/" + lampPin + '/0');

  // Answer
  answer = {
    status: 0
  };
  res.json(answer);

});
```

We start the server at the end of the file with:

```
// Start server
app.listen(3000, function () {
  console.log('Home automation system started');
});
```

We now have a server that we can use to control & monitor your home using all the modules that we deployed. However, we will now create an interface on top of this server that will allow us to easily monitor & control the whole system.

First, let's see build the graphical interface itself. It starts by creating a set of buttons to switch the appliance on or off:

```
<div class='row'>

    <div class='col-md-1'></div>
    <div class='col-md-2'>Lamp</div>
    <div class='col-md-3'>
      <button id='on' class='btn btn-block btn-primary'>On</button>
    </div>
    <div class='col-md-3'>
      <button id='off' class='btn btn-block btn-warning'>Off</button>
    </div>

  </div>
```

After that, we create two indicators for the sensor module:

```
<div class='row'>

  <div class='col-md-1'></div>
  <div class='col-md-2'>Temperature</div>
  <div class='col-md-3' id='temperature-status'></div>
  <div class='col-md-2'>Humidity</div>
  <div class='col-md-3' id='humidity-status'></div>

</div>
```

We also create an indicator for the motion sensor:

```
<div class='row'>

  <div class='col-md-1'></div>
  <div class='col-md-2'>Motion Sensor</div>
  <div class='col-md-3' id='motion-status'></div>

</div>
```

Finally, we create a field for the stream from the camera:

```
<div class='row voffset50'>
  <div class='col-md-1'></div>
  <div class='col-md-2'>Camera</div>
  <div class='col-md-7'>
    <div id="webcam">
      <noscript>
        <img src="http://192.168.0.105:8080/?action=snapshot" />
      </noscript>
    </div>
  </div>
</div>
```

Let's now see the `script.js` file that will basically make the link between the interface and the server. It starts by linking the buttons to the correct routes on the server:

```
$( "#on" ).click(function() {

    // Set lamp ON
    $.get('/on');

});
```

```
$( "#off" ).click(function() {

  // Set lamp OFF
  $.get('/off');

});
```

Then, we create a loop that will regularly update data coming from the motion sensor:

```
// Indicators
setInterval(function () {

  // Current
  $.get('/motion', function(data) {

    if (data.status == true) {
      $( "#motion-status" ).text("No Motion");
    }
    else {
      $( "#motion-status" ).text("Motion Detected");
    }

  });

}, 2000);
```

We also create another loop to update the temperature:

```
setInterval(function () {

// Temperature
  $.get('/temperature', function(data) {

    $( "#temperature-status" ).text(data.temperature);

  });

}, 2000);
```

And finally, we do the same for humidity:

```
setInterval(function () {

  // Temperature
  $.get('/humidity', function(data) {
```

```
        $( "#humidity-status" ).text(data.humidity);

    });

}, 2000);
```

Note that I also added some additional piece of required code inside the interface and as I only highlighted the most important parts here, I recommend getting the whole code from the GitHub repository of the book.

It's finally time to test the interface! Put all the code inside a folder on your computer or on another Raspberry Pi and type the following command:

```
sudo npm install request express
```

Once all the modules are installed, start the server with this command:

```
sudo node interface.js
```

Now, simply navigate to the IP address of your computer (via localhost) or to Pi on which you started this server. For example:

```
http://localhost:3000/
```

This is what you should see:

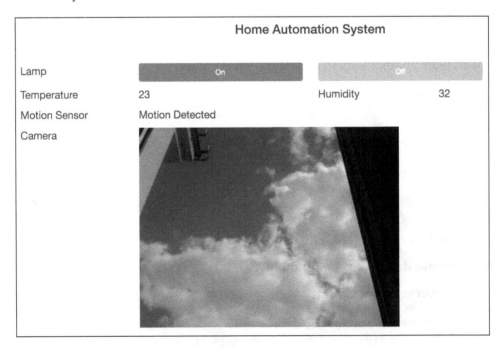

You can now try the interface; for example, using the buttons to control an appliance or by passing your hand in front of the motion sensor you should immediately see the result inside the interface. You now have a central interface for your home automation system that you can use to control your smart home!

Automating your home

Now that we have our central server running and you can use it to control & monitor your home from a single interface, we can actually define some behaviors inside the server in order to create some automation within your smart home.

As an example, we are going to automatically switch on the lamp when motion is detected by the sensor. You can imagine the scenario in which the appliance control module is connected to a lamp in your hallway and that you want it to automatically switch on whenever a movement is detected by the motion sensor.

For that, here is the code you need to add into the server code:

```
setInterval(function() {

  // Check sensor
  request("http://" + motionSensorPi + "/digital/" + motionSensorPin,
    function (error, response, body) {

      // If motion was detected
      if (body.return_value == true) {

        request("http://" + lampPi + "/digital/" + lampPin + '/1');

      }
      else if {

        request("http://" + lampPi + "/digital/" + lampPin + '/0');

      }

  });

}, 1000);
```

Let's now see what this code does exactly. We basically check the state of the motion sensor every second and then control the lamp accordingly. Note that every request is taking the correct IP address of the Raspberry Pi board to control.

You can now use the same approach to build more complex behaviors into your home, for example by linking the measurements made by the sensor module to an appliance or an electrical heater, which we already saw earlier in the book.

Accessing your home automation system from anywhere

In the last section of this chapter, we are going to learn how you can access the interface of your home automation system from anywhere in the world. This way, you will be able to monitor and even to control your home when you are not around.

For that, we are going to use a tool called Ngrok, which will allow us to access the server running on our Raspberry Pi or computer from anywhere in the world.

If like me you deployed the server on another Raspberry Pi (as my computer is switched off when I am away from home), type the following command:

```
wget https://bin.equinox.io/c/4VmDzA7iaHb/ngrok-stable-linux-arm.zip
```

This will download Ngrok on your computer. Then, unzip the file with this command:

```
unzip ngrok-stable-linux-arm.zip
```

Finally, start Ngrok using the following command:

```
./ngrok 3000
```

This will basically create a web tunnel to the web server that is running on port 3000. Inside the window that appeared on your Raspberry Pi, you should now be able to see the URL that you can use to access your Raspberry Pi from outside of your Wi-Fi network.

You can now visit this URL and you should be able to see the exact same interface as before:

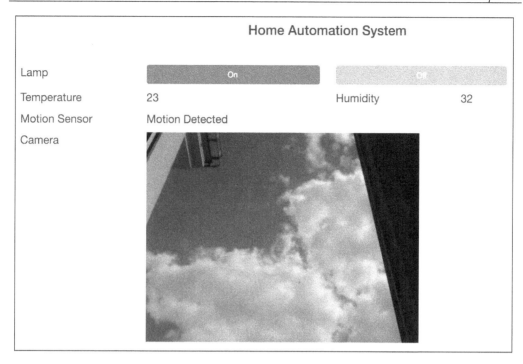

You can use this URL to access your home automation interface from anywhere in the world!

Summary

In this final chapter, we used everything we learned in the book to build a complete home automation system based on the Raspberry Pi. We used several of those boards and connected each one of it to a different type of components and then interfaced all those boards with a common interface running on your computer or on another Pi.

We also learned how to create more complex behaviors and use this central server to make the Raspberry Pi modules communicate with each other. Finally, we also saw how to access your home automation system from anywhere in the world.

You now have all the tools to transform your home into a smart home using the Raspberry Pi Zero board. I hope that this book allowed you to understand how to use this small, cheap but incredibly powerful board to automate your home. I now invite you to experiment with all the projects we saw in the book and I can't wait to see what you are going to do with it in your own home!

Index

A

alarm module
   creating 106-108
aREST framework 105
   reference link 133

C

cloud dashboard
   creating 130-133
cloud thermostat
   building 152-156
component
   testing 34-37

D

data
   monitoring, from cloud dashboard 122-129
   reading, from DHT sensor 21, 22
   remotely, accessing 24, 25
   sensor, storing 22-24
   stored, plotting 26-29
DC motor
   reference link 58
   speed, controlling 54-57
DHT11 sensor
   reference link 18
Dweet.io 123

F

Freeboard.io
   URL 125

H

HighCharts 26
home appliances
   controlling 58-62
   reference link 62
home automation system
   accessing, from anywhere 174
   components 160
   graphical interface, building 169
   graphical interface, testing 172
   hardware requisites 159
   implementing 173
   modules, building 160-163
   modules, configuring 163-166
   modules, integrating into single interface
       166-173
   software requisites 159

I

IFTTT
   URL 82
   used, for creating motion-activated lamp
       144-152